From High Peaks to Black Sea Shores: Georgia's Diverse Landscapes

Alisha

First Printing, 2024

CONTENTS

1 INTRODUCTION

[T]o that in life which tends to make these tangible substances count for most in the daily lives of a people, namely, good-will, fellowship, mutual sympathy and social intercourse among a group of individuals and families who make up a social unit, the rural community.
(Farr 2004: 11; cited from Hanifan 2016)

1.1 Initial interest to study the phenomenon

Located between the mountain ranges of the Greater and Lesser Caucasus Georgia is, with a total of 69,700 km^2 the smallest successor state of the former Soviet Union. Georgia is, however, considered as "hotspot" of biodiversity on both a European and global scale (Bedoshvili 2008; Zimina 1978):[1] Of the world's 34 so-called "biodiversity hotspots" Georgia lists two and is designated as "a haven for biodiversity" (IUCN 2012). It is a place where "[e]cosystems and landscapes change over short distances from high mountains of the Great Caucasus, and from a fairly western European climate to Mediterranean landscapes by the Black Sea and steppes in the East and South East" (Didebulidze/Plachter 2002: 89).[2] In addition to its great variety of landscapes, "Georgia is considered to be one of the centres of origin of many domesticated plant and animal species, greatly supporting the agrobiodiversity of the region" (IUCN 2012). Of the country's overall territory, mountains cover 54%, highlands make up 33%, and valleys or lowlands comprise 13% (Didebulidze/Plachter 2002: 90). Due to its diverse climate and soils Georgia's agricultural sector has a rich history (Didebulidze & Urushadze 2009: 241). The Georgian Soviet Socialist Republic was not only an important food (and minerals) supplier, but served as "center of tourism for the centralized state economy" (Slider 1995: 190). In Soviet times, agriculture generated high value export products, e.g. tea and citrus, as well as "wine, other alcoholic beverages, fresh and processed fruits and vegetables, essential oils and spices" (Didebulidze & Plachter 2002: 90). The country also "played an important role in supplying food products and minerals and as a center of tourism for the centralized state economy" (Slider 1995: 190). Production was organized either in large-scale, state-owned enterprises (*sovkhoz*) or collective farms (*kolkhoz*) that were centrally managed and controlled; rurally located families additionally cultivated 0.25 ha of state-owned land for their own production – the so-called *household plot* (Ebanoidze 2003).

---

1 The discovery of "the world's oldest wine - a vintage produced by Stone Age people 8,000 years ago" is attributed to findings in Georgia (Keys 2003; McGovern 2007).

2 Traditionally, cattle were kept in Western Georgia while sheep keeping prevailed in Eastern Georgia (Didebulidze & Plachter 2002: 97).

With the dissolution of the Soviet Union agricultural output decreased dramatically, as was the case in other republics of the former Soviet production and distribution system (WTO 2009a: 62). Agricultural production, which accounted for approximately 50% of GDP in 1991, only generated 10% of GDP in 2015. Simultaneously, the amount of people involved in agriculture actually increased. In 1990, 25% of the population were employed in the agricultural sector, whereas today more than half of the labor force is engaged (part- or full-time) in agriculture, with subsistence farming dominating (World Bank 2015b).

In a televised address in 2007, the incumbent President declared that securing property rights was "a cornerstone of our country's development" and raised objections against "those people who make allegations about the violation of property rights in Georgia:[3] soon our country will join the list of the top 20 business-friendly countries" (Civil Georgia 2007b). Though multiple cases of property evictions later became publicly known (TI Georgia 2010, 2012a, 2012b), the President's promise did come true: Georgia climbed from 112th rank in 2005 to a top Doing Business performer classified by the World Bank in 2008 (World Bank 2009: 16). In fact, thanks to a newly established land register, Georgia ranks first in property registration worldwide (World Bank 2015a).

My decision to focus on land privatization in Georgia was prompted by the findings of an explorative research stay in July 2011[4], which was initially directed toward analyzing overgrazed pasture land and a rapid increase of exported sheep beginning in 2010. Further evidence gathered during my fieldwork shifted my focus to examining the current state of property rights in Georgia, the work of donor agencies and diverse patterns of land governance, influencing both the export of sheep and overgrazing. It came to my attention that the export of sheep from Georgia is based on the efforts of one individual who, during Soviet times, came from a rather well off-family and was later financially supported by the Millennium Challenge Corporation (MCC) and the United States Agency for International Development (USAID), followed by organizational support by the Georgian government.[5] Subsequently, a rising number of press releases were reporting on land evictions at the same time as a new property registration scheme was be-

[3] As many of the sources used for this thesis appear to have been written by non-native English speakers, there are a number of deviations from common English usage in the quotations used throughout. Rather than unduly altering them or indicating my awareness of these language issues with the conventional *sic*, I will leave them as is, unless they seem likely to lead to misunderstanding.

[4] As a master's student of the Social Sciences at the Ivane Javakhishvili Tbilisi State University, Georgia (2006–2008), I was able to gain insight into the local context and became familiar with the agricultural sector as an independent consultant investigating "Farmers' Organisations in the Development of Agriculture in the South Caucasus" for the Evangelischer Entwicklungsdienst (EED) in collaboration with the Caucasus Research Resource Centers (CRRC), from June to October 2007.

[5] Interview with an expert on sheep herding and exporting (July 13, 2011, Tbilisi).

ing launched, meaning that formalization of property rights was underway. Although property formalization and its effects have been studied extensively in Africa, for example, research on this topic in the Southern Caucasus has remained quite limited.[6] Thus, I concluded, analysis of Georgia's case could fill an important gap in the literature and add to our understanding of the diversity of approaches to, and consequences, of land privatization.

1.2 Previous studies

The demise of the Soviet Union together with the fall of the Berlin Wall spurred an interest in institutional change in the so-called countries in transition (Blanchard 1997; Swinnen 1997: 2; Arrow 2000; Barrell et al. 2000; Macours & Swinnen 2000; Roland 2000; Svejnar 2002; Swinnen & Heinegg 2002; Hagedorn 2004; Swinnen & Rozelle 2006; Beckmann & Hagedorn 2007; Bromley & Yao 2007; Gomulka 2007; Spoor 2009). Work in this domain generally examines a variety of effects stemming from the pace and scope of reforms but also tends to emphasize that initial conditions matter, and context is one of the main determinants of the paths that transitions take. Moreover, Starr (1988: 6) emphasizes that "[p]rivatization is a fuzzy concept that evokes sharp political reactions. It covers a great range of ideas and policies, varying from the eminently reasonable to the wildly impractical". It is, accordingly, understood as "a policy movement and as a process that show every sign of reconstituting major institutional domains of contemporary society" (ibid. 1988: 6). Research on Georgia's transition, especially with respect to the privatization of agricultural land, is limited. Starr's reflection on the challenges of the transition process though mirrors well how the allocation of property rights have shaken politicians and farmers to their very foundations during these first years.

In Georgia, the first reforms intending to privatize land were led by the Republic's first President, Zviad Gamsakhurdia (1990–1992), in the aftermath of a coup d'état which was accompanied by inter-ethnic wars in the breakaway regions of South Ossetia and Abkhazia as well as civil unrest in Samegrelo, which led to an unstable domestic political environment until 1995. Georgia's less than favorable initial conditions were aggravated by the fact that "constant changes in laws and institutions […] led to the often violent settlement of disputes and to people retreating deeper into a private world of patronage, family support, and illegal meth-

[6] This line of thought is supported by the fact that searching for literature on the Web of Science (latest search on July 24, 2017) brought only four results for "land privatization Georgia", of which only two refer to the topic in question, whereas one article deals with agro-food processing and the other with metal contamination of agricultural soils. Meanwhile, a search for "land individualization" generated only one result. The three relevant sources are presented below.

ods of survival" (Jones 2000: 49). As a result, the rural population has been relying on traditional methods of production because, as previous field work has shown, the bonding nature of interactions between peasants provides a much more secure environment than changing production and risking their familiar and habitual setting (Buschmann 2008). Georgia's GDP dropped to 20% of its prewar level till mid-1990s (Christophe 2004a: 90), whereas the share of agriculture reached 64.4% in 1994 (Kegel 2003: 148).

Based upon this disparity, the process of distributing land has been characterized as having been "designed to be incomplete" (ibid. 2003: 12), in view of the fact that "[t]he land reform process was incomplete and gave discretion to the local elite and the managers of the state and collective farms to influence [...] implementation" (ibid.: 13). In particular, "people who were empowered during the previous system tried to preserve their network and to benefit from privatization" (Bezemer & Davis 2003: 9). Moreover, "the initial land reform legislation did not provide legal procedures for the introduction of full and enforceable private property rights" (ibid.). Within this context, Lerman (2004a) summarizes the main obstacles surrounding land transactions in contemporary Georgia as follows: First, bureaucratic complexity, namely high transaction costs due to complicated bureaucratic procedures, inadequate computerization, staff shortages and preemptive purchasing rights for local populations; second, administrative restrictions and an absence of alternative employment opportunities; and hence, third, economic behavioral choices among rural populations: Land is regarded as a safety net which decreases incentives to sell it, whereas leasing is viewed as being equally less attractive, as a result of "weak contract enforcement" (ibid. 2004: 71). According to USAID (2010: 7), the issue of bureaucratic obstacles toward land transactions has been solved: By 2010 it is said that the registration of land has been made "relatively inexpensive, simple, fast, and effective" (ibid. 2010a: 7), due to the simplification and digitalization of the registration process (ibid.: 8).

The latter, the so-called formalization of property rights, meaning official registration of rights to immobile property located within a particular cadaster had been introduced by law in 1999 whereas the set-up of a cadaster by the National Agency of Public Registry (NAPR) was not completed before 2008. The registration process though is viewed critically, however, due to its high cost and organizational flaws. First, it costs at least 153 Georgian Lari (GEL) [~ 62 USD] "for registering one piece of land up to 500 square meters" (ibid.).[7] Second, holdings are generally fragmented, with agricultural land plots of 2–4 parcels per household on average (cf. Lerman 2004a). Third, remote areas were never subject to land surveying and, hence, have been excluded from the cadastral system, so "requests to register property located in these zones have been denied" (TI Georgia 2012a).

[7] 1 Georgian Lari (GEL) = 0.404984 USD (September 20, 2017).

Christophe (2004a) analyzed institutional arrangements on the local, regional and central levels in Kutaisi, one of Georgia's former Western administrative units, showing that frequent changes of rules and constant amalgamation of the formal and informal sectors are – in contrast to most other theoretical accounts attempting to explain the emergence of corrupt networks – not signs of an often-proposed "weak state" that is not capable of stability and enforcing its own rules. Rather, Christophe illustrates the complementary nature of a system that uses formal institutions to secure access to resources through informal institutional arrangements, where the production of uncertainty serves as the very means to secure power (Mardin 1969).[8] The ruling elite has applied two primary techniques to maintain (relatively efficient) political control over economic resources (Christophe 2004a: 94–95): First, through steady encroachment on formally recognized rights to private ownership, via the state's ability to break and change its own formulated rules at any time, and, second, through the state's potential to selectively and randomly regulate and sanction as a means to discipline economic competitors. The ruling elite is, thus, not comprised of an oligarchy that colonizes state agencies but, rather, a bureaucratic elite that possesses legal means of violence and manipulates the state's regulative capacity to prevent the creation of an independent economic counterweight (ibid.: 95). Especially vis-à-vis donors, Georgian state officials typically present simulations of political reforms to secure further access to financial resources (2004a: 73–74).

According to Christophe, in Georgia *official* rules neither give any legally binding guarantees nor provide clear instructions for officially recognized practices; meanwhile, *unofficial* rules may be linked to networks of political supporters that either provide exclusive access to the market or might be used to block potential opponents from such access (ibid.: 97). In particular, Christophe underlines the *social* origin of power as a traditional Georgian characteristic, where personal relationships are not the consequence but, rather, a precondition for accessing lucrative posts (2004a: 60). Moreover, seen through the lense of path dependency, in contrast to the other Soviet states, the Georgian Socialist Republic exhibited an extraordinarily high degree of institutionalized corruption, where informality and criminality were always closely related (Christophe 2004a: 59; Turmanidze 2001).[10] In addition, the limited capacity of the socialist state to penetrate the rural populace not only led to the exclusion of the rural population from many correc-

8 In total, Christophe identifies five patterns that have shaped and maintain the Georgian power structure: First, creation of uncertainty; second, manipulation of conflicts; third, privatization of risks; fourth, destroying of interpersonal trust; and, fifth, preemptive cooptation of potential competitive elites (2004a: 89–179).

10 Often cited is the fact that the productivity of privately cultivated fruit and vegetable plots was five times higher than in state-owned kolkhozes (ibid.: 59). Thus, even though official numbers showed an average income per household, Georgians possessed the most cars and houses in comparison to other Soviet republics (ibid.: 89).

tive measures by the Socialist regime but also to the preservation of many informal traditional practices. For example, Soviet brigades were comprised of relatives, and their payment schemes were not in line with official Soviet rules but, rather, according to the social status of their members (namely, position within the state hierarchy) and the size of land that was brought in by a household to the joint production site (ibid.: 60).

Christophe thus identifies three sources for the institutionalization of contemporary agricultural property-related behavior in Georgia: former socialist state structures, neo-patrimonial clientelistic networks, and the survival of traditional, clientelistic arrangements (ibid.). It is exactly the fusion and changing applicability of these rule sets which makes reasonable (economic) foresight virtually impossible (ibid.: 100). In contrast, these conditions provide room for maneuver and favor the interests of insiders, while outsiders are kept out and potential competitors eliminated. Additionally, some sectors exhibit agglomerations of ambivalent rules, such as vis-à-vis local governance (Turmanidze 2002), while others, foremost trade and transport, are based on a detailed body of rules and regulations that can hardly be followed (ibid.: 88). The latent non-institutionalization of specific rules serves as an instrument to spread uncertainty and facilitate control (ibid.: 121). Especially from Turmanidze's work on the legal basis of land privatization in the 1990s, we can see that details on how to organize the distributional process administratively were explicitly left out in a way that ultimately privileged a specific clientele (ibid.: 135). In this vein, she stresses the outstanding capacity of the Georgian state apparatus to adapt constantly to changing environments (ibid.: 83).

When Georgia regained its independence, as in other republics of the Soviet production and distribution system, agricultural output decreased dramatically (WTO 2009b: 62). Seeing that Georgia was the sole provider of tea, wine and citrus fruits within the Soviet system, it consequently suffered from losing its former monopoly rents (Christophe 2004a: 89). The Government of Georgia (GoG), led by Eduard Shevardnadze (1992–2003), thereafter received heavy financial assistance from international donors, mainly the European Union, World Bank and International Monetary Fund (IMF; (Papava 2003). Moreover, Georgia did not liberalize its trade sector to any significant degree nor did it manage to diversify its export markets but, instead, continued to rely on agrarian production. Privatization measures, on the other hand, failed to create competition among banks and other formally state-owned enterprises (Gylfason & Hochreiter 2009: 366). Though prices were stabilized and inflation kept under control, corruption and related issues remained a major challenge up to the change of political leadership in 2003 (ibid.).

Given these conditions, the immediate post-Soviet epoch has been characterized as being rife with "corruption, organized crime, nepotism, human rights abuses and ethnic strife" (Jones 2000: 42), up to the transfer of power to Mikheil Saakashvili (2004–2013). The reign of the latter though – initially accompanied by massive support from Western donors for his successful steps toward "reducing bureaucracy, combating petty corruption, improving tax collection, service delivery and infrastructure" (Mitchell 2009: 171) – resulted in a severe imbalance of governmental branches (ICG 2007). A massive power transfer toward the executive was set in motion (Reisner & Kvatchadze 2005: 13), with the judiciary, which lacked any experience in modern commercial law, becoming dependent upon it (EBRD 2006: 19). Saakashvili's reign was further criticized for a lack of access to information on political processes and lack of transparency vis-à-vis the political decision-making processes. In addition, reforms devoted to extending municipalities led to a decline of finances and, consequently, a lack of access to various services on the local and regional levels (TI Georgia 2008). In particular, requirements to levy property and land taxes became extremely difficult to enforce, due to an intransparent property transfer process that took place from the central to the local levels (TI Georgia 2008: 3).

Consequently, reforms aiming to increase the role of self-governing bodies on the local level failed. As Reisner and Kvatchadze (2005: 13) propose, "the conflict about the distribution of, and access to[,] scarce resources especially in rural areas has been intensified, since it creates opportunities of self-enrichment for a few in, or those connected to[,] central positions". Critiques of the lack of "transparency, accountability and credible investigations into disturbing cases of official abuses" (ICG 2007: i) mounted, culminating in public protests in 2007, which resulted in massive use of force "against peaceful demonstrators, the violent closure of a private television station and the imposition of emergency rule [which] brought a halt to hitherto unquestioning Western support of the Georgian leadership" (ibid. 2007: i). Thus, in the aftermath of Saakashvili's reign, critical calls grew louder regarding the role of donors and non-governmental organizations (NGOs). A paper by Chatham House argues that, as a result of the post-Soviet transition process in Georgia,

> a rather elitist non-profit-organization sector emerged, which focused on professional consulting and service provision. [...] The elitist nature of NGOs is largely attributable to the fact that their main sources of funding are foreign. Western money allows NGOs to attract talent, but their full-time employees are more comfortable networking with Western embassies and various state agencies than holding town hall consultations and engaging with citizens. (Lutsevych 2013: 4)

Accordingly, the funding practices of donors toward NGOs has resulted in a so-called NGO-cracy, with a negative impact on Georgia's civil society:

> Western funded organizations are not anchored in society and constitute a form of 'NGO-cracy': a system where professional NGO leaders use access to domestic policy-makers and Western donors to influence public policies without having a constituency in society. (ibid.: 4)

Although bottom-up voices from civil society representatives are rather weak, trust in the church remains relatively high, leading to "a strong social outreach allow[ing] religious institutions to play a role in the development of civil society" (ibid.: 9). As an expression of this power, "[o]ne of the largest demonstrations in recent years in Georgia was a protest march in Tbilisi by thousands of people, led by the priesthood, against the law on religious minorities" (ibid.: 9).[11] Consequently, "[t]he weakness of civil society not only renders citizens helpless to prevent backsliding by ruling elites, it also allows those holding power to commit abuses". In 2012, the newly elected Prime Minister, Bidsina Ivanishvili, opened the way for passing new laws to improve the independence of the judiciary (TI Georgia 2013) and set the course for "a dramatic change in the government's spending priorities" (TI Georgia 2014b). Whether any changes actually resulted from this that led to more support for the Georgian agricultural sector will be treated below.

In the late 1990s, Lerman reported on the outcomes of land distribution in Georgia and their impact on agricultural production (Lerman 1996, 1999), indicating that small-scale farming was constrained by a lack of agricultural machinery, an inability of small-scale farms to specialize, and the high prices of inputs (Revishvili & Kinnucan 2004: 56). In a later survey, Lerman (2004) compared land reforms in the three South-Caucasian Republics – Armenia, Azerbaijan and Georgia – which all exhibited high levels of agricultural production and employment in relation to overall GDP, especially in view of the relative scarcity of land in those countries. These three states applied similar approaches in favor of land distribution to individual households but, Lerman stresses that, unlike the land-quality-oriented provisioning in Armenia and Azerbaijan, allocations in Georgia were granted based solely on local availability of land (Lerman 2004b: 66), and soil quality or the availability of water were not taken into consideration.

Cemovich (2001) provides an overview of the land-reform process in Georgia, with formalization of ownership seen as the final step. The author highlights the lack of a legal basis to land ownership prior to 1999 and outlines the effects created by obstacles to transfering state property to individual users. Land was transacted on the basis of trust and lacked any documentation, leading to gaps between official records and actual land holdings (Cemovich 2001: 85). Meanwhile, Kegel (2003) stresses that average farm size in Georgia varies greatly regionally, ranging from 0.39 to 0.96 ha, with densely populated villages having even lower average sizes (Kegel 2003: 150). Bezemer and Davis (2003) illustrate how earlier managers of kolkhozes and sovkhozes, as well as other members of the local elite, have benefited from the transition process, namely by keeping control over "privatized" assets, either through family ties or links to local officials and their networks (Bezemer & Davis 2003: 9; Swinnen 1997). Another avenue for profiting from

[11] The support of a rather homogenous, not open, society by the church can further be seen in the fact that "93 per cent of respondents in Georgia [...] said they would not like to have homosexuals as neighbours" (Lutsevych 2013: 9).

the reforms was provided by legal changes in the taxation system (Bezemer & Davis 2003: 9–10). No land tax was levied until 1997, but the list of taxes to be paid to the local bureaucracy became steadily longer starting in 1999, with the result that some farmers began paying bribes to local tax-collection officials, which amounted to less than the taxes themselves, whereas others decided not to register their land to avoid further tax payments (ibid. 2003: 10).[12]

A study by Ebanoidze (2003) gives an overview of the decade-long process of transferring property rights to land from the state to individual households, providing a synopsis of the legal framework, land administration and management structures as well as the involvement of many donors, such as USAID, GIZ and the Word Bank, that were engaged in supporting emergence of the land market (see below and chapter 4.3). The author critically assesses the low share of privatized land and the subsequent small, fragmented plots that resulted from land distribution and the lack of coordination among the various donors to finalize the process of land registration.[13] Moreover, it is emphasized that the "functions of administering and disposing of particular state properties and of land registration function are not differentiated" (Ebanoidze 2003: 138). Moreover, the author stresses the overlapping of local and state agencies assigned with land administration, management and policy development (ibid. 203: 138).

The share of state-owned agricultural land that was leased out in the early 2000s – about half of the total land under state ownership – is analyzed by Tsomaia (2003a), revealing that a quarter of the total land owned by the state in the early 2000s was leased by only a handful of individuals or legal entities (Tsomaia 2003: 8). On the other hand, Tsomaia highlights the short leasing periods of ten years or less granted in 1996, whose continuation after contract end was still uncertain at the time of the enquiry. Salukvadze (2006) summarizes the outcomes of the first phase of land privatization and shows that, by 1999, about 75% of agricultural land was still under state ownership, whereas 55% of arable, 68% of perennial, 29% of hay and 5% of pasture land had been privatized (Salukvadze 2006: 5). He also illustrates the first years of implementing the title registration (or land administration) system and the lack of co-ordination among the multiple donor organizations, resulting in system incompatibilities due to different and competing technical approaches (ibid. 2006: 8).

---

[12] Sikor and Mueller (2009: 1309) reveal the challenges posed by state-led top-down initiatives, which "cause land reform programs to miss out important developments on the ground and fail to enlist support from relevant actors. Reliance on bureaucratic modalities hinders the adaptation of state action to tenure arrangements and authority relations on the ground".

[13] Land fragmentation in Georgia is identified by Lerman (2005: 1) in terms of *land use*, which he calls "farm fragmentation", in contrast to fragmentation of *land ownership*, as was the result of restitution in e.g. the Czech Republic or Bulgaria, where plots were divided among several heirs (cf. Sklenicka & Hladík 2009; Dirimanova 2005). His survey results stress that "land productivity declines as farm fragmentation increases" (Lerman 2005: 4).

Gvaramia (2013) gives a detailed breakdown of land privatization in Georgia and focuses on the installation of a 21-km long state-border corridor where privatization of land was prohibited (Gvaramia 2013: 6). But even though Article 14 of the Constitution of Georgia grants the principle of equality "regardless of race, color, language, sex, religion, political and other opinions, national, ethnic and social belonging, origin, property and title or place of residence" (1995), the limitations were mainly placed upon "the Kvemo Kartli and Samtskhe-Javakheti regions owing, most likely, to the large non-ethnic Georgian local populations" (Gvaramia 2013: 6). Also another study emphasizes both the distribution as well as the leasing out of state land to rural dwellers as being a highly intransparent processes, for local governments did not provide sufficient information to local residents (GTZ/CIPDD 2006: 7). As a consequence, "land was rented out primarily in the form of large lots to firms and private persons" (ibid. 2006: 7) not belonging to the region, whereas local farmers subleased the land, at times to several intermediaries; meanwhile, other leaseholders did not rent out the land but "hired local peasants as day labor" (ibid.: 7). These cases reveal unequal treatment of Georgians and ethnic non-Georgians with regard to receiving land, which were finally resolved through land allocations in 1994: The border corridor was reduced to a 5-km swath in 1998 and, although restrictions were placed on land use to preserve border security, such as prohibitions against constructing certain kinds of structures therein, farming activities and land ownership within the zone were allowed (ibid.: 7). Gvaramia (2003) furthermore outlines the legal changes resulting from the approval and operation of the Civil Code of Georgia from 1997, which introduced mandatory property registration. The first law that allowed the registration of rights to land and other fixed came into effect in 1997. In 2004, the National Agency for Property Registration (NAPR) was launched, which replaced the State Department for Land Management (SDLM).[14] From 2004–2006 the data was transferred from the SDLM to NAPR. In 2005, the Parliament of Georgia adopted the Law on Registration of Rights on Fixed Assets. In 2008 the Law on Public Registry eventually "regulates the activities of the Public Registry as well as all issues related to the registration of rights" (Gvaramia 2013: 8); from 2009 the NAPR was finally in charge of property registrations under the Ministry of Justice (MoJ; USAID 2011b: 5).

Legal proof of ownership obtained during the initial reform process from 1992 on was the so-called handover document, issued by a local self-governing body

[14] From 1998 to 2004, the SDLM was responsible for implementation of agricultural land reforms and registering rights in fixed assets (not land). Before 1998, the the Bureau for Technical Inventory (BTI) was responsible for registering buildings. With the liquidation of the SDLM, the NAPR became engaged in registering property rights in land as its primary task, though it also grew in other competences and is now responsible for registering any kind of ownership in the public and private/business domain (Gvaramia 2013: 8).

(*sakrebulo*) (Gvaramia 2013: 8).[15] According to estimates, 30–35% of the newly created land "owners" did not obtain any document, but registration of ownership at NAPR is still possible by applying to the Archive Unit of the relevant municipality that keeps records.[16] In the aftermath of the the reforms, the GoG initiated a systematic land cadaster with the land's registrations free-of-charge (ibid.: 9). Collaboration with the USAID-funded Land Market Development project (1999–2005) "ensured a systematic cadastre measuring of up to 2.5 ha land parcels and their registration into the Public Registry" (ibid.). Seeing that the latter included a registration fee of GEL 50 [USD ~ 20], in 2012 the Public Registry began implementing a project for "the systematic primary registration of agricultural lands and the specifics of their borders" (ibid.), through which all agricultural land in Georgia was to be registered but the populace would not be charged for the surveying and registration services involved. Yet, the survey work registered by NAPR was reportedly marked by inaccuracy, with "survey and other errors still aris[ing] with some frequency. The extent of the undiscovered and uncorrected errors is not known, and estimates of current registrars range [from] 10% to over 50% of cases, depending on the area" (USAID 2011b: 14). This situation has been further aggravated, because the resulting median-sized farms of 0.75 ha are split between two or three plots (Lerman 2005: 1), and resurveying of individual parcels is required together with registration of ownership (Bechtolsheim et al. 2012: 6), but "the current rule is that the costs of correcting the survey work has now been shifted to applicants for registration in the new cadastre, which may delay completion of the new cadastre and be burdensome for lower income land owners" (ibid. 2012: 6). As a result, based on the latest estimates, only 20–30% of agricultural land has been registered thus far (USAID 2013: 7). A subsequent hitch to land registration also arose:

> The associated costs are regarded as relatively high, especially in rural areas (with low plot values). As a result, land sales often take place solely on the basis of a transfer certificate [...] and the transaction does not get officially recorded within the land register. Such informal transfers of property ultimately will complicate proof of ownership and facilitate respective disputes, especially over the long term. (KfW 2011: 4)

Hence, even though primary registration and measurement of land boundaries was to be carried out free-of-charge, high actual costs being passed on to farmers – to

---

[15] In some cases, the SDLM certified the origin of land rights stemming from the initial reform era (ibid. 2013: 8).

[16] The process is regulated by the Law of Georgia on Recognition of Property Rights of the parcels of Land Possessed (Used) by Natural Persons and Legal Entities under Private Law (2007). Lawful possession applies to the possession of land and adjoining buildings before 2007, together with holding one of the following documents: "certification characteristic (issued by the Technical Bureau Archive of the Public Registry), excerpt from the Household Book or the list of land distribution (to be found at the district archives), book of the gardener (document certifying membership of collective gardening) or certification of assets from district archive".

register their landed property, plus entailed surveying costs – seem to have counteracted the intended effect. Summing up this contradictory situation and its results, the Georgian newspaper *Kviris Palitra* declared that "land privatization, which has to be completed by May 1, 2011, is suspended at this stage, because the government works on new rules and prices" (DWVG 2011a). To the best of my knowledge, no study has yet analyzed the institutional genesis of property rights in agricultural land and its impact on agricultural production in Georgia. Thus, the following study is intended to fill this niche.

1.3 Land privatization examined through the lenses of Public Choice and New Institutional Economics

The present thesis is premised on the ideas of the New Institutional Economics (NIE), a rather novel research paradigm in the social sciences (Williamson 2000: 596).[17] Both classical political economists, associated with Adam Smith or John Stuart Mill, and early neoclassical economists, such as Paul Samuelson or John Maynard Keynes, "tend to disregard the institutional framework within which economic activity takes place, and the ongoing exchange relationships that constitute that activity in the first place" (Boettke & Marciano 2015: 54). The strength of this more recent stream of thought and the reason for its actual application is its focus on (the economics of) institutions and the costs of exchange as a measure for understanding and endogenizing alternative ways of economic organization and allocation of resources (Eggertsson 1990: 10; Williamson 1975; Coase 1960).[18] The NIE permits relaxation of several axioms of the neoclassical economics' agenda, particularly the rational-choice model as well as the assumptions of costless exchange and full information (Eggertsson 1990: 6–7; Stigler 1961). By bringing in the effects of uncertainty in economic exchange – one of the key features of the Georgian transition process with regard to its objectives, the outcomes and implications of the interactions between the various reforms undertaken, as well as associated political constraints (and opportunities) producing

[17] This stream of thought links insights from e.g. Public Choice Theory, Property Rights Theory and Transaction Cost Economics that apply the neoclassical logic based on cost–benefit calculations systematically to the level of institutions (Hagedorn 1992: 192). As economic choices over scarce resources target minimizing both production as well as transaction costs, institutional arrangements evolve that attempt to keep both low, in comparison to alternative forms of economic organization.

[18] The neo-classical criteria of comparing economic outcomes in terms of "efficiency" has proven problematic, however, for "the economist can never say that one social situation is more 'efficient' than another. This judgment is beyond his range of competence" (Buchanan 1959: 137–138). Others propose, "[w]hat matters is not efficiency, but *efficiency for whom*?", (Bromley 1989b: 4). The same applies to endeavors intended to find "optimal" processes of exchange, where output is normative in nature (Eggertsson 1990: 10).

both winners and losers (Roland 2000: 12–13) – NIE can take into account pervasive measurement problems encountered and actors' inherent drive toward reducing transaction costs (Eggertsson 1990: 27). In the Georgian case, measurements costs have arisen as a consequence of the underlying property rights structure, the function of the courts and the legal system, "and the complementary development of voluntary organizations and norms" (North 1990b: 64).[19] As this study investigates transaction and transition costs under diverse institutional arrangements shaping property rights to land in Georgia, I have drawn on NIE concepts to provide an analytical focus on the diverse contractual arrangements that have emerged, highlighting the nature of the rights transferred and the conditions defined for realizing exchange (ibid. 1990: 45). Central here is a concern for economic logic and the competitive nature of contractual arrangements, meaning trying to grasp why one contractual form dominates another (ibid.: 53).[20] The analysis is thus of a positive nature, focusing on the economic impacts associated with alternative institutional arrangements, in accord with Coase's "Institutional Premise":

> Without some knowledge of what would be achieved with alternative institutional arrangements, it is impossible to choose sensibly among them. We therefore need a theoretical system capable of analyzing the effects of changes in these arrangements. (Coase 1988: 30)

Instead of following neoclassical postulates of individuals with stable preferences and a consistent profit maximization motive, the NIE turns to Herbert Simon's behavioral model of individuals aspiring toward "satisficing" under *bounded rationality* (Simon 1972, 1983). According to Simon (1983), human behavior is constrained by a limited capacity to gather and process information adequately; objectives may vary over time and are shaped by evolving mental processes (Eggertsson 1990: 9).[21] Accordingly, "[r]ationality of the sort described by the behavioral model doesn't optimize, of course. Nor does it even guarantee that our decisions will be consistent" (Simon 1983: 23). Simon's concept also incorporates the impact of emotions in the analysis of institutional change, which is a vital constituent of land ownership in Georgia, as upheavals and demonstrations against the free-marketization of privately held agrarian and communal pasture land to non-locals in the period from 2010 to 2013 has shown. The search for and

---

[19] Besides, measurement costs specifically arise in cases of measuring quality, seeing that asymmetric information may lead to Akerlof (1970) market of "lemons", where goods of low-quality drive high-quality goods out of the market (Eggertsson 1990: 196).

[20] For example, a particular form of insurance is non-specialization, applied when the costs of transacting and uncertainty are high, the latter defined as consequence of a problem's complexity and an individual's cognitive capacity to solve a given problem (North 1990: 34).

[21] Simon (1983) bases his concept on evidence gathered by Amos Tversky and colleagues, especially when dealing with situations characterized by uncertainty (see e.g. Tversky & Kahneman (1974).

comparison of alternatives is, according to Simon (1983: 28–29), grounded on recognition and intuition which, in turn, is determined by knowledge gained through past experiences.

Assuming that "the basic structure of property rights is determined by the state" (Eggertsson 1990: 79), the role of the state within NIE is to impact resource allocation through initial distribution or reallocation of rights or by blocking their reassignment: "Any redefinition of the structure of property rights by the state has wealth effects involving both winners and losers" (ibid. 1990: 40). Endogenizing "the framework of rules that govern social intercourse" (Boettke & Marciano 2015: 62) became both a theoretical agenda and an empirical challenge within Public Choice Theory, also termed the New Political Economy, which was developed from Rational Choice Theories and emerged as a response to a rather static, centralized and hierarchical theoretical tradition in public administration (Ostrom & Ostrom 1971: 203). As emphasized by the Ostroms, "Simon elucidated some of the accepted administrative principles and demonstrated the lack of logical coherence among them" (ibid. 1971: 204). Simon (1964) thus established an administrative theory based for the comparison of alternative institutional arrangments that evaluates the relative efficiency of available options (ibid. 1971: 204).

Primarily associated with works of Black (1948), Arrow (1951), Riker (1962) and Niskanen (1971), the role of government, policies and attendant bureaucracy have been assessd on the basis of economic principles, leading "to a theory of the failure of political processes" (Mitchell 1988: 107) based on "perverse incentives embedded in rules of collective choice" (ibid. 1988: 107). Members of government, whether in coalitions, teams or as single individuals, are considered to act rationally by formulating policies and serving interest groups with the aim of gaining or keeping themselves in office (Downs 1957). Just as market actors compete for sales, government members and office-seekers compete for votes in electoral processes, including seeking support from agricultural interest groups (Hagedorn 1991, 1994). As Hagedorn proposes, "[t]he number and the unity of interest groups is a mirror of the policy process" (1994: 414). Establishment of a few large interest groups, as is the case in Germany or Britain, where a single farmers' association maintains a monopoly position, is supported via incentives by the state bureaucracy, which prefers to negotiate with only a small number of organizations rather than with individual splinter groups (ibid. 1994: 414).

Developed during the 1950s and 1960s among political scientists and economists, three geographic locations in the US – Rochester, Virginia and Bloomington – became the cradles of this research field that centers on how social choices evolve (Mitchell 1988: 102–104), as detailed in the following.

Rochester school

Researchers based at the University of Rochester (New York) established elements of positive political science via mathematical explorations, including elaborating formal tools and applying behavioral models to voting and the effects of decision rules.[22] Arrow (1951), for example, showed that, in a two-party electoral system, no stable outcome is to be expected under repeated majority voting, while Black (1958) assessed coalition formation and party strategies in his study on voting in committees, and Coleman (1966) analyzed strategic coalitions, namely the effectiveness of vote-trading and log-rolling, by applying game theory.

Virginian school

The Virginian stream of institutional thought, notably associated with the work of economists such as Buchanan (1959b, 1958), Tullock (1967) and Olson (1965), rejects the tenets of orthodox neoclassical economics – such as centering on equilibrium outcomes or maximizing social welfare functions (Boettke & Marciano 2015: 54) – but is rooted in the contractual tradition and, regarding "politics as exchange" (ibid. 2015: 56), critically assesses the role of political processes (Mitchell 1988: 106–107).[23] By focusing on the logic of constitutional decision-making, (the origin of) legislative institutions and "critical institutional elements of collective choice" (ibid. 1988: 104), these authors set the foundation for a constitutional political economy (Boettke & Marciano 2015: 57; Buchanan 1990). For example, in a "study of non-market decision-making" (Ostrom & Ostrom 1971: 203), Buchanan & Tullock (1962), who also focused on log-rolling and vote-trading from a constitutional point of view, distinguished and demonstrated the tradeoffs between the costs of compromise and representation (external costs) and the costs of decision-making (opportunity costs) under alternative constitutional rule sets. By applying economic principles to political processes, "[b]oth Coase and Tullock recognize limits to economies of scale in bureaucratic organization."[24] The presumption of self-interested, rational decision-makers leads to "[u]tility maximizing individuals [who] seek out exchanges, not conflict" (Mitchell 1988: 107). In contrast to this new premise of individuals operating under bounded rationality, rational choice and, later, public choice approaches assume (1) rational decision-making, (2) maximization strategies of actors that depend on (3) levels of information possessed, which put individuals in a position of *certainty*, *risk* or *uncertainty* (Ostrom & Ostrom 1971: 205). The *condition of certainty* applies to situations where "an individual knows all available strategies,

[22] Researchers originating from the Rochester school have also been located at Carnegie Mellon University, the California Institute of Technology and the University of Texas (Mitchell 1988: 102).

[23] Researchers who developed the Virginian approach to Public Choice were located at the Thomas Jefferson Center for Studies in Political Economy at the University of Virginia in Charlottesvile, the Center for the Study of Public Choice at the Virginia Polytechnic Institute in Blacksburg, and the George Mason University in Fairfax (Mitchell 1988: 101).

[24] In comparison to Coase's (1937) theory on the nature of the firm, Tullock (1965) views limitations on the size of the firm being similar to those affecting large bureaucratic organizations: Though long-term employment contracts lower decision-making costs, the management and productive capacity of too many employees outweighs the benefits of their integration (Ostrom & Ostrom 1971: 209).

[...] each strategy is known to lead invariably to only one specific outcome, and [...] the individual knows his own preferences for each outcome" (ibid. 1971: 205). In a *condition of risk*, an individual is aware of the stragegies available to her, but "[a]ny particular strategy may lead to a number of potential outcomes, and the individual is assumed to know the probability of each outcome" (ibid.: 206). Here, a decision depends on an individual's preferences for different outcomes and the probability of their respective occurrence and, hence, becomes a weighting process where "an individual may adopt mixed strategies in an effort to obtain the highest level of outcomes over a series of decisions in the long run" (ibid.). In a *condition of uncertainty*, it is assumed that an individual may be either aware of all possible strategies and outcomes "but lacks knowledge about the probabilities with which a strategy may lead to an outcome, or [...] may not know all strategies or all outcomes which actually exist" (ibid.). Hence, in conditions of uncertainty, only estimations regarding the consequnces of available strategies can be made, and it can be assumed that the "individual learns about states of affairs as he develops and tests strategies". Furthermore, "[i]ndividuals who learn may adopt a series of diverse strategies as they attempt to reduce the level of uncertainty in which they are operating (ibid.).

Accordingly, Tullock (1965) in his studies on public administration depicts the figure of the "economic man", "an ambitious public employee who seeks to advance his career opportunities for promotions within the bureaucracy" (Ostrom & Ostrom 1971: 209). Consequently, in their efforts to please superiors, such career-oriented public servants will tend to forward favorable information but generally withhold less favorable information, leading to a situation of distorted information that "will diminish control and generate expectations which diverge from events sustained by actions. [...] Large-scale bureaucracies will, thus, become error prone and cumbersome in adapting to rapidly changing conditions" (ibid. 1971: 209). The democratic policy process thus results in "inequity, inefficiency, and coercion" (Mitchell 1988: 107), seeing that "both politicians and bureaucrats face perverse incentives [so that] publicly-offered goods end up being over-supplied and costly" (ibid. 1988: 108). The political market, then, is not only shaped by "property rights, contracting and credible commitment" (North 1990a: 356–357) but also by "imperfect information, subjective models, and high transaction costs" (ibid. 1990a: 357). Under the assumption that "public policy reflects a series of bargains among various interests" (Weingast & Marshall 1988: 133), legislative institutions are seen as serving to facilitate exchange among different interest groups – in accordance with their bargaining strength – while legislators themselves exchange rights over specific jurisdictions that can allow them to exert influence over the political process. Weingast and Marshall emphasize that, "[l]ike market institutions, legislative institutions reflect two key components: the goals and the preferences of individuals, here legislators seeking reelection from their constituents, and the transaction costs that are induced by imperfect information, opportunism, and other agency problems" (ibid. 1988: 134). But, unlike market transactions, legislative institutions depend on enforcement mechanisms in the form of durability of coalitions (ibid. 1988: 143–144). The bidding for and trading of votes (similar to market exchange) is complemented by legislators who "bid for seats on committees associated with rights to policy areas valuable for their reelection" (ibid.: 148). Thus, the goal is to gain influence through "the property rights established over the agenda mechanism, that is the means by which alternatives arise for votes" (ibid.:157).

Moreover, politicians are led by ideological stereotypes, resulting in so-called imperfect (subjective) models that guide their action; the interplay of different policies to be enacted makes it difficult to precisely envisage particular outcomes. The outcomes produced are thusly influenced by the legislator's choice of representation, which subsequently lead to policy outcomes that favor particular constituents (North 1990a: 363). From the perspective of examining transactions, North also underlines how measurement and enforcement costs not only arise between legislators and their constituents but also between legislators and the agents who enact policies (ibid. 1990a: 362–363). Meanwhile, evaluation of policies by constituents is generally complex, due to high measurement and enforcement costs that tend to provide low payoffs for constituents seeking to acquire information (ibid. 1990a: 361). The political market is, therefore, marked by imperfect information for both legislators and conituents, resulting in high transaction costs and, consequently, less "effective", meaning costlier solutions for constituents (ibid.: 361; cf. Krueger 1988). Hence, political outcomes typically do not match envisioned goals. Another key factor, proposes North, is that "in most legislation redistribution is either concealed or a by-product of other objectives" (North 1990a: 357).

Khan (1995) particularly addresses the "political costs" or "transition costs" which are inherent to solutions seeking to remedy institutionally caused "state failures", stressing that "[i]t is necessary to be explicit about these costs and recognize that their incidence is not equal or inevitable" (Khan 1995: 71). He distinguishes between the output of an existing set of institutional arrangements and the complementary process of attempting to bring about institutional change to improve them, with "outcomes [being] compared in terms of a chosen criterion such as utility, net output or growth" (ibid.: 73).[25] Khan also identifies two kinds of structural failure: Type I and Type II. To him, Type I "*occurs if a particular formal institutional structure results in lower net benefits for society compared to an alternative structure*" (ibid. 1995: 73; italics in original). Such benefits, apparently decreasing due to state intervention that has set incentives contrary to the common good, are conceptualized within NIE under the categories of transaction costs and rent-seeking. Transaction costs, a field of research that stems from organizational theory and is fundamental to NIE in a broader sense, are the costs of exchange that arise both ex ante as well as ex post from imperfect contractual

[25] The nature of institutional change is explained in detail in chapter 3.3.2.

arrangements and, consequently, impact the allocation of resources (Williamson 1975; Coase 1937). As such, "[t]ransaction costs are detrimental for social net benefits because they prevent gainful transactions from occurring which might otherwise have taken place" (Khan 1995: 75). As a result, Khan proposes, "Coase's insight that transaction costs differ across institutions underlies the NIE analysis of Type I failure" (ibid. 1995: 75). Meanwhile, the concept of rent-seeking (Tullock 1967; Krueger 1974; Posner 1975; Buchanan 1980), which is associated with research in trade theory and the New Political Economy more generally, is based on the assumption that "state-created rents create incentives for agents to leave productive activities for so-called unproductive ones to try and acquire credentials which give access to the rents" (Khan 1995: 74).[26] One key effect of rent-seeking is "the use of productive resources in unproductive activities" (ibid. 1995: 74). According to Khan, "[m]oves towards laissez faire are predicted to reduce the incidence of rent-seeking and hence Type I failure" (ibid.: 74). On the other side of the coin, Type II state failure occurs during the process of institutional change and indicates failures of transition (ibid.). When it arises, there is a need "to compare alternative paths to a better structure [... as] specified by theory or observation" (ibid.: 73). As formulated by Khan, it "*occurs when the process for changing the structure of institutions attains a lower cumulative set of net benefits for society compared to an alternative process over a given period*" (ibid.: 74; italics in original). In fact, he notes, "the existing process of change may be increasing the magnitude of Type I failure" (ibid.).

To investigate Type II failure, NIE relies on higher-level transaction costs, such as North's political transaction costs or the costs of organizing collective action, to understand the institutional determinants of processes of change (North 1990a). This typically involves comparing alternative paths, meaning "[i]f an alternative process could have carried out a transition to a better structure or carried it out faster, the cumulative difference in net benefits over a period of time gives a measure of Type II failure" (Khan 1995: 73–74).

Bloomington school

By incorporating into its perspective philosophical dillemmas, "entailing justice, liberty, and equality" (Mitchell 1988: 108), the aim of the Bloomington school has been to overcome the forced separation "between philosophy, politics, and economics" (Boettke & Marciano 2015: 55). Under the auspices of Buchanan in Virginia, "[p]ublic choice emerged more or less as an empirical research program in public economics, and as a methodological critique of welfare economics and public policy economics" (ibid. 2015: 61). Whereas that critique was primarily

[26] This is because, as supporters of these models argue, "the cost of state intervention was more than the traditional deadweight welfare losses associated with the divergence of prices from marginal costs. [...] The withdrawal of resources from productive uses continues till the expected marginal return to a factor from productive and unproductive activities is equalized" (Khan 1995: 74).

directed against the growth of government, the focus of scholars centered around the University of Indiana at Bloomington shifted from the size (scale) to the scope of government (ibid. 2015: 62), dealing generally with constitutional provisions (Mitchell 1988: 108). As Mitchell explains their relationship, "Bloomington has taken cues from and is in basic accord with the Virginians. On the efficiencies to be gained from decentralization and on the importance and efficiency of the price mechanism and competitive markets the two schools are in complete accord" (ibid. 1988: 112; cf. Ostrom et al. 1961). Work associated with Bloomington, notably by political scientists who founded the Workshop of Political Theory and Policy Analysis, emphasizes "study[ing] real constitution making from the bottom-up, as opposed to from the top-down as is often done in idealized theory" (Boettke & Marciano 2015: 62). In this vein, "Vincent Ostrom has chosen, for the most part, to consider public choice as a philosophical problem, while Elinor Ostrom has devoted her career to the empirical testing of propositions from public choice and political science dealing with the provision of public services" (Mitchell 1988: 110). The Workshop has dealt foremost with the "conceptualization of public goods as the type of event associated with the output of public agencies" (Ostrom & Ostrom 1971: 105) and fostered "a long-enduring interest in federalism, natural resources" as well as "the methodological status of political theory and public choice" (ibid. 1988: 110). Their work has been appreciated for the merit of being able "to convert the generalized message" of evaluating the output of government "into more operational language" (ibid.: 112; Ostrom 1976, 1975). The Bloomington framework of inquiry is comprised of institutional variables, namely constitutional and post-constitutional choice sets that define the so-called rules of the game (Ostrom 2005); (Ostrom et al. 1994). With this, "the Ostroms have made the study of constitutional order a touchstone of their scholarship" (Mitchell 1988: 110). In addition to developing frameworks for "theoretical foundation of research in policy analysis" (Ostrom 2011: 7), to be used as "a general language for analyzing and testing hypotheses about behavior in diverse situations at multiple levels of analysis" (Ostrom 2007: 21), Bloomington-based research has in more recent years centered on managing common-pool resources and "how rules, physical and material conditions, and attributes of community affect the structure of action arenas, the incentives that individuals face, and the resulting outcomes" (ibid. 2007: 21;(Ostrom 2008),).

Based on the economics of property rights literature that flourished in the 1960s (Williamson 2000: 598–599; Demsetz 1967; Alchian 1965; Coase 1960), for Bloomington researchers ownership and its entailed decision-making rights are understood as economic-choice variables which fulfill specific economic functions in that they structure incentives, and "the crucial aspect of property rights is [seen as] their function as a mechanism for social control of individual behavior" (Dahlman 1980: 3). Considered in this way, property rights encourage indviduals "to behave in certain ways, and to avoid behaving in certain others. […] The tenet is that, by influencing incentives, property rights can be used to reduce or avoid such costs of transacting, if they are designed and enforced properly" (ibid. 1980: 3). A change in long-established contractual arrangements, such as leasing agreements among farmers for example, may turn out to be a long process, "particularly when there is a lack of experience with arrangements that would be best suited to a new situation" (Eggertsson 1990: 55). It is, hence, important to underline that "the ability to adjust successfully to changes in the environment is not distributed evenly among individuals" (ibid. 1990: 57–58), depending for example on availability of information, budget constraints or the ability of individuals to alter the status-quo structure of property rights "in order to minimize their personal cost of unexpected exogenous changes" (ibid.: 58), as newly imposed imports, a ban on the sale of new products, or declaring particular contractual forms illegal. Such processes may result in a re-distribution of property rights and a shift in the land-labor ratio so as to transform a group of property owners into farm workers. Thus, "property rights, far from being dangerously flexible, often lag behind changes in the environment and act as brakes on economic development and growth" (ibid. 79). In fact, "[i]t is the analysis of such crippling inflexibility that perhaps constitute the most interesting task ahead for the NIE", proposes Eggertsson (ibid.: 79). From this perspective, the structure of property rights "reflects the preferences and constraints of those who control the state" (ibid.: 79).

All in all, the NIE provides a positive understanding of the role of the state which, at its best, can enforce contracts through its sanctioning power and, thus, reduce the costs of transacting (ibid.: 46). Research advanced by NIE places greater emphasis on empirical observation than mathematical calculation (Eggertsson 1990: 6) and, when evaluating alternatives to the status-quo structure of entitlements, the goal is "to devise practical arrangements which will correct defects in one part of the system without causing more serious harm in other parts" (Coase 1960: 34).

1.4 Research questions and aims of the study

The main *research question* pursued in the present thesis is the following: How did political reforms targeting land privatization in the Republic of Georgia affect land governance and tenure, and what have their effects on agricultural production been since the break-up of the Soviet Union? The research question has an explorative purpose, to generate insights about the effects of this induced change toward privatizing land as well as to develop tentative hypotheses on the topic (Rubin & Babbie 2010: 41). More particularly, my research has been guided by the following sub-questions: (i) How did these reforms create winners and losers? (ii) Did people oppose change in the past or try to coalesce against it? If yes, how? And, finally, (iii) how did the (non-)allocation of property rights affect agricultural production and, hence, the distribution of wealth among the rural populace of Georgia?

I suggest that studying land privatization and its subsequent effects is worthwhile, in part, because individual land ownership and the marketization of land is a novelty in Georgia's rich history, where initially feudalism and, more recently, collectivization dominated the modes of agricultural production. Hence, this study centers on the creation of property rights and their substance over time, in an environment where no exclusive rights to factors of production had existed before. Given this newly emergent situation, in an environment organized by households in rather homogeneous village structures predominantly inhabited by Georgian and minority Armenian or Azeri communities, comparatively examining the differing impacts of land reform on land tenure and agricultural production seems particularly promising. In addition, because Georgia's agrarian production is determined by various climate zones and differs from one region to the other, my analysis seeks to reveal what kinds of diverse effects these conditions may have on land tenure throughout the Republic. The present work also examines the differing effects stemming from the "wholesale" approach toward land privatization initially pushed by the government versus the "incremental" market approach promoted by international donor organizations vis-à-vis the formalization of land tenure to spur agricultural production. The end goal of my analysis is to weigh the net effects of Georgian land privatization by identifying who benefited and who suffered from the reforms and, consequently, hypothesizing what the potential future impacts of this process on land tenure and agricultural production may be.

I consider this study's *contribution* to be both theoretical and practical: first, understanding the impacts of institutional change via written land governance arrangements on tenure formalization and, second, investigating their effects on actual agricultural production in Georgia. Accordingly, the study shifts attention from an initial focus on policy makers, donor agencies, legal advisors, other practitioners and researchers to, later, evaluating the empirical outcomes of land allocation and tenure. In particular, I scrutinize the final process of property registration of agricultural land in Georgia in terms of the following issues and goals:

- characterizing the livelihoods of rural populations, viewed as the most vulnerable part of the society, who require the crafting of durable, sustainable and equitable solutions to the challenges that have resulted from the landed-property reforms;
- promoting the organizational capacity of land-governance institutions in the long run, on the basis of well-founded knowledge, seeking to ensure sustainable land use, support investments and encourage improved agricultural production;
- informing, engaging and communicating with the scientific community, donor agencies and practitioners on the specific challenges of formalizing land ownership in Georgia;
- and, finally, understanding the impacts of the potentially different forms of induced institutional change toward land privatization in Georgia.

My *working hypothesis* has been the following: Political reforms tackling land privatization have benefited those close or belonging to informed political circles who are now better off, as measured in terms of high agricultural output and secured allocated property rights. Meanwhile, it is hardly to be expected that "outsiders", meaning those not regularly involved in political matters, including the poor and minorities, have benefitted from the reforms. I base these claims on the following line of reasoning: Land reforms commenced with the distribution of state-owned agricultural land in such a manner that state officials and those close to the government found their way to the best pieces of the cake. At the same time, evidence suggests that members of the political elite were allocated more land, whereas minorities got less and were subsequently hired as a cheap labor force. Moreover, the poorer layers of the society have barely means to keep, rather than to expand production and, hence, the windows of opportunities for the better-off and more well-informed in society have remained open. This particularly applies to formerly leased state-owned agricultural land and its subsequent acquisition as well as to land sold (conditionally) via auctions, as it was reported by Gvaramia (2013) that in 2003 more than half of Georgia's leased state land was contracted by a few individuals for speculative purposes.

1.5 Organization and structure

Though advanced inverted on the ground, the present study's seven chapters are organized analytically and, hence, not chronologically. In line with an abductive research approach, the empirical work has largely preceded the selection of theory, while in an iterative arpproach theoretical views have been continually reviewed, revised and adapted throughout the research process based on empirical knowledge gained.

Chapter One has provided an introduction to the topic, disclosing my interest in exploring land privatization in Georgia and providing reasons for doing so through the lense of New Institutional Economics (NIE) and Public Choice theories, as they enable examination of contractual relationships, frameworks of rules governing transactions and their subsequent distribution of costs and benefits, as well as assessment of alternative institutional arrangements. My literature review of previous studies on the object of research has revealed that research dealing with land privatization in Georgia is quite scarce and, to the best of my knowledge, the kind of institutional analysis being undertaken here has not been carried out yet. My research questions have formulated my intention to critically scrutinize both the means and outcomes of the Georgian land reforms; thereafter, I have presented the purpose and expected theoretical and practical contributions of the study as well as my working hypothesis.

Chapter Two centers on the study's tools, namely the methods and research design employed. It introduces the abductive research approach that I have applied together with grounded theory, triangulation and a mixed-methods approach. In addition to steadily searching for and including primary and secondary literature on the topic, an exploratory quantitative survey conducted in Georgia by the Centre for Social Studies (CSS) in 2012 preceeded my qualitative field work there in 2013 and 2014.[27] As the CSS survey was focused on the role of social capital among Georgian villagers (CSS 2012), the group of involved researchers allowed me to include five questions related to land tenure and agricultural production in the questionnaire. Hence, the section describes the survey's stratified sampling method, used for the selection of respondents, as well as analytical techniques applied using SPSS. Based on prior results stemming from my own previous research (Buschmann 2008) or earlier empirical work by others (see chapter 1.2), I first formulate tentative hypotheses on the impact of socio-economic factors (independent variables) on the choice of whether to register land (dependent variable), then test them with SPSS on the basis of frequency distributions and the degree of correlation to the registration of land, and finally depict the results graphically to reveal important linkages. The subsequent section presents the methods used to undertake the field work, including the selecting of study sites and respondents. Then, qualitative analysis – based on expert interviews and focus groups within the framework of two case studies, evaluated using Atlas.ti – seeks to better comprehend the links between socio-economic features and land registration and, consequently, identify the winners and losers of the reforms and why.

Chapter Three provides the theoretical and analytical foundations for the study on land privatization and begins by defining institutions as norms and conventions, working rules and property relations (entitlements), providing insights regarding their implications for the distribution of costs and benefits in contractual relationships. The next section deals with organizational forms for governing transactions in socio-ecological contexts and, thus, focuses on the transaction as the unit of analysis. Against the background of the recently introduced property rights formalization in Georgia, organizational forms are subsequently discussed in more detail, with a view to the assessment of polycentric systems for the provision of public goods in general as well as land administration and registration of ownership in particular.

---

[27] The Centre for Social Studies (CSS) is a think tank located in Georgia's capital, Tbilisi, which is supported by the Academic Swiss Caucasus Net (ASCN), a program initiated by the Swiss-based Gebert Rüf Foundation, that supports the social sciences and the humanities in (Eastern Europe and) the South Caucasus and promotes cooperation among Caucasian and Swiss scholars, see https://www.grstiftung.ch/en/area-activity/closed-areas/osteuropa2.html. CSS' research project was financed by the Academic Swiss Caucasus Net, in cooperation with the University of Fribourg (see http://www.ascn.ch/en/research/Completed-Projects/Completed-Projects-Georgia.html).

Chapter Four presents an overview of socio-economic conditions at the time of Georgia's independence, portraying the diverse periods that define the process of land privatization in Georgia from 1992 to 2015, namely state-led reforms, market-based reforms and eventually donor-led tenure formalization.

Chapter Five presents my quantitative and qualitative findings. The quantitative results provide a general picture of tenure formalization and low land-registration rates, including correlations between tenure formalization and socio-economic variables related to household/farm characteristics as well as traits of cohesion and collective action in the CSS-survey villages. My qualitative findings are based upon charting the develop of the legal framework for land privatization in Georgia, complemented by an institutional analysis of four action situations using the Institutions of Sustainability (IoS) framework to analyze the land privatization process in depth. This part of the study begins by presenting the characteristics of agricultural production in both Eastern and Western Georgia, focusing on wine and hazelnut production. After that follows an assessment of the properties of landed resources in Georgia, where scarcity prevails, making income stemming from both such assets (stock) and their flows (yield/land rent) highly valuable future benefit streams particularly worth competing for. The respective institutional and organizational forms that govern land-related transactions are then examined within each of the four action situations.

Chapter Six discusses and draws conclusions from the study's results. The chapter is sub-divided into two parts: conclusions and implications. First, conclusions are drawn based upon the study's theoretical and empirical underpinnings (see chapters Three and Four) and then, derived from that, theoretical and political implications regarding land privatization and ownership (in Georgia) as well as implications for future research are formulated.

The Appendices are comprised of a brief historical review emphasizing the political divisions between Eastern and Western Georgia (Appendix I), an overview of my field research during 2013 and 2014 (Appendix II) as well as a description of the theoretical and empirical assumptions underlying SPSS data analysis (Appendix III).

2 TOOLS OF THE GAME: METHODS AND RESEARCH DESIGN

Florence Nightingale revolutionized hospital practice by insisting: Whatever else hospitals do they should not spread disease. And so, the idea [...] is that whatever else jurisprudence does it should not spread confusion.
(Cohen 1954: 376)

2.1 Methods

2.1.1 Scientific logic and reasoning

The present study is based on Grounded Theory and the logic of abduction. The latter, in the form popularized by Strauss and Corbin (Strauss & Corbin 1990), is considered a method for collecting and processing data by using "theoretical pre-knowledge [which] flows into the data' s interpretation" (Reichertz 2010). In accord with the abductive logic of Peirce,

> the research is laid out in such a way that new hypotheses can and do appear at every level, that the interpretation of the data is not finalized at an early stage but that new codes, categories, and theories can be developed and redeveloped if necessary. (ibid.: 2010: 7)

An *abductive* approach is an alternative mode of inference to the other two classical forms that either derive conclusions "by perfect logical processes from well-defined premises" (Arthur 1994: 406), in the case of *inductive* reasoning, or by looking for patterns and simplifying a problem "by using these to construct temporary internal models or hypotheses or schemata to work with" (ibid. 1994: 406), in a *deductive* manner. The abductive method of processing data involves "assembling or discovering, on the basis of an interpretation of collected data, such combinations of features for which there is no appropriate explanation or rule in the store of knowledge that already exists" (Reichertz 2009: 15). Subsequently, the task becomes to discover or invent a new rule:

> Here one has decided (with whatever degree of awareness and for whatever reason) no longer to adhere to the conventional view of things. This way of creating a new "type" (the relationship of a typical new combination of features) is a creative outcome which engenders a new idea. This kind of association is not obligatory, and is indeed rather risky. (ibid. 2009: 13)

As a result, abductive proceedings "seek some (new) order, but they do not aim at the construction of [just] *any* order, but at the discovery of an order which *fits* the surprising facts; or, more precisely, which solves the practical problems that arise from these" (ibid. 2009: 23; italics in original).

2.1.2 Grounded Theory

To complement the abductive approach, the present study is based on Grounded Theory which was originally developed by Glaser & Strauss (1967a) as a systematic technique for qualitative (and specifically interpretative) studies and, thus, seeks to facilitate deeper understanding of social phenomena (Strauss 1994: 19). The method was influenced by classical works of the American Pragmatists, including Charles Sanders Pierce, William James and John Dewey as well as Herbert Blumer's "symbolic interactionism" (Mead 1934), which takes a constructivist perspective on the social nature of people's behavior.[28] The method was initially applied in the area of medical care, more precisely for a field study on family members coping with dying patients in hospitals (Glaser & Strauss 1965, 1968). Today, it is used by researchers in many areas of the social sciences, pedagogy and health care, due to its general way of analyzing data that does not depend on a single discipline (Strauss 1994: 19).

The technique enables to elaborate conceptual meanings of amounts of data (Charmaz 1996: 27). Moreover, the approach combines the research process with the theoretical development and hence "blur[s] the often rigid boundaries between data collection and data analysis phases of research" (ibid. 1996: 28). It is essentially based on four aspects inherently associated with its approach, which seek to facilitate and enrich its application, namely (1) case-specificity; (2) socio-scientific interpretation as an art; (3) linking everyday life's and scientific thinking; and (4) openness of socio-scientific conceptualization (Strauss 1994: 11–14).

(1) Case-specificity: A "case" is understood as an autonomous "action unit" that has its own background and is subject to its own logic (ibid.: 12). The typical purpose of reconstructing a case is scientific theory formation, so that primary data is not used to simply illustrate a pre-set theory but, rather, theoretical ideas are constantly reviewed in light of a specific case, contrasted and adjusted accordingly (ibid.). Developing a theory is understood as a form of interaction between a researcher and the study object – "something appears *as* something *for* somebody" (ibid.; italics in original) – depending on the epistemological interests of the researcher (ibid.). In this way, depiction of reality possesses a particular structure and certain limits and, thus, forms a case (ibid.). Referring to the example of Glaser and Strauss (1965), dealing with dying patients and the reactions of their families and friends, each patient forms a single case; but a case might also consist of a hospital, a family or a person or even deal with social connections (social worlds), such as the milieu of an AIDS victim or membership in a sports club (Strauss 1994: 12).

[28] It is assumed that people act vis-à-vis objects (understood as social objects) on the basis of the various meanings given to the latter, a result of social interaction, which are formulated and modified in an interpretative process. Social structures of meaning are, thus, a human product and subject to continuous change (Strauss 1994: 30).

(2) Socio-scientific interpretation as an art: The socio-scientific interpretation of reality (to form a theory) can be compared with creating an artwork (ibid. 1994: 13). Both share two distinct features, as they both (try to) start with an unknown, *unbiased view* and, subsequently, *create reality* as an ensuing product of confronting reality artistically or scientifically (ibid.). Referring again to the example of Glaser and Strauss (1965), an adequate socio-scientific approach to investigating how family members are treated by hospital employees following a family member's death needs to have an unbiased view on a singular case against the background of being theoretically informed (Strauss 1994: 14). By studying research participants in an open but profound way, Charmaz (1996: 30) emphasizes that "[t]he researcher seeks to learn how they construct their experience through their actions, intentions, beliefs and feelings".

(3) Continuity of everyday and scientific thinking: A researcher's everyday-life and scientific understandings of human interaction do not differ structurally, though the latter strives to attain a greater degree of explicitness, since a researcher is (usually) free of compulsion to take action, the so-called *Privileg der Handlungsentlastung* (roughly, exemption from practical action; ibid. 1994: 14). Thus, for Grounded Theory, everyday knowledge is regarded as an indispensable resource for the scientific process" (ibid.: 13) and is used systematically together with professional knowledge via "theoretical sampling" (ibid.: 13–14). Understanding derived from the experiences of everyday life can yield sociologically meaningful, and even crucial theoretical contributions relevant for those working in the respective professional sphere. Thus, investigating a case may also serve as a starting point for collaboration between scientists and practitioners (ibid: 14.).

(4) Openness of socio-scientific conceptualization: Scientific knowledge, just as reality, has a processual character that is in steady conflict between determination and surprise, and whose meaning is therefore subject to (social) change (ibid.). In line with Grounded Theory, research, including the production of theories, terms, concepts and categories, requires steady verification to gauge the accuracy of its applicability for each particular scientific investigation. The purpose of analyzing qualitative data is to generate theory (ibid.: 29), which itself might end up being at diverse levels to make the investigation and its outcomes more explicit (ibid.: 28). At the lowest level, the outcome might be descriptive, simply the reproduction of a respondent's information, for example. Meanwhile, at the highest, namely complex and systematic, level, the intention could be to develop a theory in the form of an interpretation, as Grounded Theory already postulates that "*[a]nalyzing* is equal to *interpreting* the data" (ibid.: 28; italics in original). In the end, theories might be located on each potential level, with different ranges of assertion, and possibly linked with other theories (ibid.: 29). However, theories are to be generated within closest possible proximity to the data by researchers who understand themselves as instruments for developing Grounded Theory (ibid.: 31). Professional literature needs to be a synthesis of the researcher's personal experience and contextual knowledge (ibid.: 36–37): Focusing on the triad

data collection, coding and writing memos is seen as a means to control inadequate integration of personal attitudes and opinions (ibid.: 37). Scientific theories are established, first, by formulating hypotheses that are based on ideas or assumptions; second, implications are derived from already existing (systems of) hypotheses; and, third, the process is finalized by verifying a hypothesis, evaluating whether it can be confirmed fully, partially or not at all (ibid.). It is important during this process for researchers to reflect on the implications of collected data and experience with former studies or previously acquired contextual knowledge to develop (even vague) hypotheses (ibid.: 38).

Broadly speaking, the main phases after data collection during Grounded Theory analysis are comprised of (A) encoding the data, (B) comparative analysis, (C) theoretical sampling and (D) integration of the results through coherent theoretical depiction (ibid.: 44–48). (A) The encoding process begins by reflecting on the collected data in connection with contextual knowledge and developing generative questions. Then links are established between investigated concepts and, – by establishing a code for each specific question, being it a part of or a concept of its own – a kind of dense "theory" begins to emerge. (B) Preliminary ideas on the nature of the relationships between concepts are verified (or not) through further investigation and the resulting new data and new codes, which are under review throughout the process. As the data becomes integrated in a more coherent manner during the course of this examination, the core of the ensuing theory takes shape. (C) Key categories are then worked out, which are intended to link together all other categories. Throughout the analysis, but especially at this stage, theoretical ideas are developed with the aid of theory memos, which are reviewed and sorted from time to time and may, thus, yield new ideas. As the study progresses, these theory memos become more sharply focused. During these analytical operations – namely data collection, coding and writing of theory memos – it is important to be aware of the *temporary* and *relational* aspects of this triad (ibid.: 46), as the researcher moves between coding and writing memos, collecting new data, refreshing memos and sorting codes anew. In contrast to other qualitative methods, the use of previously existing data is desired and is generally allowed to be included during the enquiry. (D) If the resulting integration of the data is deemed to lack sufficient theoretical density, a return to the analytical triad may be necessary, depending on the explanatory power of the codes and memos, the topic(s) the researcher wishes to stress as well as the experience of the latter in terms of research and scientific publications. The purpose of this interative process of analyzing data is, according to Charnaz, to finally make the analysis both more abstract and more concrete (Charmaz 1996: 47).

2.1.3 Data analysis by the use of Atlas.ti

The qualitative analysis (see chapter 5.2) is realized by the software Atlas.ti (version 6.2.15), which enables data processing in accordance with Grounded Theory.

In short, the following steps depict the analysis by the use of Atlas.ti (ibid. 2002: 66; cf. Tesch 1990):

1. Reading through primary data, i.e. interviews with experts and focus group discusssions, apply significant units (*codes)* on textual segments and, based on these, aggregate *families* that work as categories.[29] A *code* "is a technical term from the analytical procedure and signifies a named concept. (…) The differentiated concepts are known as categories" (Böhm 2004: 271). During the coding process, the researcher makes use of the background knowledge that she or he has about the general and particular context (ibid. 204: 271).
2. Systemizing the data by steady comparisons "to build and refine categories, to define conceptual similarities, and to discover patterns" (Smit 2002: 66).
3. Analyzing is finally "a process of resolving data into its constituent components, to reveal its characteristic elements and structure" (Dey 1993: 30).

Whereas the coding of data is "central both to grounded theory and to most of the programmes developed specifically for qualitative analysis" (Smit 2002: 69), the advantage of using a software as Atlas.ti is the systemic use of technical support to construct "networks of code categories" (Smit 2002: 68).

2.1.4 Triangulation and mixed methods approach

Investigating the institutional change of property rights over agricultural land in Georgia is an interdisciplinary task which not only seeks to tackle the changing legal framework of land ownership but also the resulting changed interactions within the agrarian economy. Consequently, the study presented here is based on triangulation, a mixed method approach for combining methods to examine the various dimensions of the same phenomenon (Denzin 1970: 297), incorporating a number of perspectives and broaching different aspects of the issues at hand (Jakob 2001: 2; Flick 1995: 433).[30] In particular, the present study uses *across-method* as well as *data triangulation* (Jakob 2001: 2–3; see figure 2-1 below).

---

[29] Primary data in the present case was collected in extensive field visits by taking notes and supported by digital audio recording (to secure accuracy).

[30] The term *triangulation* is used by qualitative researchers to refer to the integration of quantitative and qualitative data; it is a metaphor derived from the fields of navigation and land surveying related to the determination of geographical locations by means of using two known points to find a third (Prein et al. 1993: 12). Although the concept is ancient in origin, Campbell & Fiske (1959) are among the first who attempted to validate their measurements with other instruments and measurements to verify and evaluate their results (Prein et al. 1993: 12). Denzin (1970, 1978) pursues the idea of linking diverse methods to validate outcomes, even though he was criticized for his approach (see e.g. Fielding & Fielding (1986), or Flick (1992); a summary of the critical discourse is provided by Prein et al. (1993). A literature review on the use and justification for a mixed-method approach

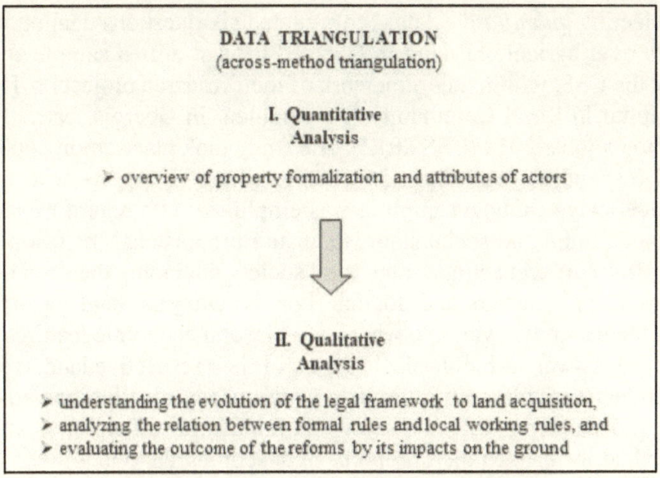

DATA TRIANGULATION
(across-method triangulation)

I. Quantitative Analysis

- overview of property formalization and attributes of actors

II. Qualitative Analysis

- understanding the evolution of the legal framework to land acquisition,
- analyzing the relation between formal rules and local working rules, and
- evaluating the outcome of the reforms by its impacts on the ground

Figure 2-1: Data triangulation (own graphic)

Across-method – a sub-type of the general methodological triangulation – seeks to increase a study's validity by applying a variety of quantitative and qualitative research techniques to produce two distinct data pools that are analyzed respectively by different techniques – in contrast to within-method triangulation, which uses different methods to analyze the same data. The latter, *data triangulation*, uses different times, locations and/or respondents for the collection of data and fits well to the application of Grounded Theory (ibid. 2001: 3; cf. Glaser & Strauss 1967b).[31] As the use of multiple sources of evidence is warranted "for the valid and reliable collection of case study data" (Theesfeld 2004: 258), both across-method and data triangulation are applied here.

The aim of the first part of my own quantitative analysis was to obtain an overview of the outcomes of property formalization in Georgia, identify the attributes of the rural population in general, and, more particularly, the socio-economic features that might have an influence on the formalization of land ownership. The subsequent qualitative analysis was intended to more deeply comprehend the reasons why specific features have emerged by (1) studying the evolution of the legal framework for land acquisition, (2) analyzing the relationships between formal rules and local working rules, and (3) evaluating the outcomes of the reforms by closely examining their impacts on the ground. Both approaches are presented briefly below.

---

is presented by Bryman (2006: 104–107), and an epistemological discussion on triangulation is provided by Olsen (2004).

31 Denzin (1978: 301) divides triangulation into two further forms: *investigator* triangulation (data collection by different kinds of observers, e.g. men and women) and *theory* triangulation (testing different theories to evaluate the problem from other perspectives).

For collecting *quantitative* data, I formulated six questions dealing with property rights over agricultural land in Georgia for a stratified sample survey conducted by the CSS, within the framework of their research project on The Role of Social Capital in Rural Community Development in Georgia, carried out from February to October 2012 (CSS 2012). The study took place among 600 respondents – 30 respondents each from 20 villages across Georgia – who were interviewed face-to-face. Quota sampling was employed with regard to age, gender, economic condition and social status to ensure proportional inclusion of people stemming from different strata of the rural society, including the so-called village intelligentsia, e.g. teachers and doctors. For the analysis, land registration was treated as the dependent variable whereas social and economic features of the actors involved, as well as biophysical aspects of their respective land, were chosen as independent variables. With the help of frequency distributions among these variables, it became possible to display initial indicators of common attributes of those who had formalized their property rights compared to those who had not.

As my overall analysis deals with induced institutional change, the second part of the investigation focused on the evolution of the legal framework, the relationships between formal rules and local working rules and the outcomes of the reforms by analyzing their impacts on the ground. Thus, after a formal legal analysis was conducted to gain an understanding of the actual legal framework, qualitative analysis concentrated on farmers, producers and processors in sectors that depend on long-term approval of their property rights regimes and which represent two of Georgia's leading agrarian export segments: hazelnuts and wine. These products also serve as good examples for gaining representative results, I propose, since hazelnut cultivation is located in Western Georgia, while wine production is concentrated in Eastern Georgia. Hence, focus groups were conducted among small-scale farmers in three villages each in East and West Georgia to investigate in-depth their motives for registering or not registering their property. Moreover, three companies from each of the two sectors were studied with regard to the nature of their property rights and their modes of operation. These methods were applied in the hope of obtaining a clear picture of the present state of the land-tenure regimes under which rural inhabitants in Georgia operate and the characteristics of their land. The study was rounded out by conducting expert interviews with, for example, (local) government representatives, employees of the National Agency of Public Registry, donor organizations, such as USAID, as well as employees of the Association for Protection of Landowners' Rights, Georgian Young Lawyers' Association and Transparency International Georgia.

2.2 Explorative quantitative survey

The CSS questionnaire mentioned above focused on the following topics:[32] (A) social and economic characteristics of each village, including changes perceived in the village, economic conditions there in relation to other villages, and its social structure with regard to minorities, outsiders and (non-)religious groups; (B) life within the village community, including communication, cooperation and unwritten rules among villagers; (T) trust, namely vis-à-vis outsiders and among native villagers; (R) respondent's place in the community in terms of relationship to other villagers, membership in formal organizations or discussion of problems with fellow villagers; (V) respondent's profile and value orientation, such as preferences vis-à-vis work and living; (F) respondent's family situation, including household size and economic conditions, involvement of the family in agricultural work and husbandry; (D) demographic information, such as gender, nationality and religion; and (M) sources of and interest in information, uncluding use of television, radio or internet and kinds of news sought.

In addition to these topics, taken as a basis for beginning to understand the characteristics of the actors involved and their everyday organization of agricultural production, CSS allowed me to include the following six questions regarding land ownership and land use of agricultural land in the the family-related section of the survey:[33]

(F4) How many hectares of land does your family own?
(F5) Is this land registered at the public registry?
(F6) If yes, when did you register the land?
(F7) If not, why did you not register the land?
(F8) How many hectares of land does your family lease?
(F9) What share of the land plots that you have do you use for agricultural puroposes?

2.2.1 Stratified sampling among Georgian farmers

The selection of the study sites is based on information that CSS obtained during the first phase of its research, after which 30 respondents were chosen from each of selected 20 villages, some of which were located near major roads whereas others were relatively hard to access.[38] Further parameters for selection of the villages sought to account for ethnicity, namely villages dominated by Georgians, a

[32] According to M. Muskhelishvili (personal communication, March 23, 2012). The questionnaire can be obtained on request.

[33] Special thanks to Marine Muskhelisvhili, the CSS team and the Academic Swiss Caucasus Net for allowing me to include these six questions (F4–F9).

[38] According to M. Muskhelishvili (personal communication, March 23, 2012), the following villages were included in the analysis: Vaka, Aghaian, QorTa, Nasakirali, Qoreti, Zoti,

minority (typically of Armenian or Azeri origin) and mixed populated villages; religion (Christian orthodox or Muslim); and economic conditions. The sample population was selected by using quota sampling, CSS' researchers did not claim to be striving for representative results: "Given [the] limited number of respondents for each village, quota sampling was considered to be the best possible option to insure inclusion of different segments of villagers."[39] In addition, the technique of stratified sampling was also used, as it allows for a sample to be representative on specific characteristics, e.g. education, as well as to represent various key subgroups, e.g. teachers or farmers (Teddlie & Yu 2007: 79).

2.2.2 Data analysis by the use of SPSS

Before the analysis began, tentative hypotheses were formulated on the impact of socio-economic factors (independent variables) on responses to question (F5): the choice to register land or not (dependent variable; for an overview of these assumptions, see annex III). These propositions were intended to not only function as guiding ideas throughout the investigation but also to disclose intuitively presumed relationships and, hence, indicate "what is relevant and what is irrelevant" (Burns & Burns 2008: 212). The propositions were either based on empirical findings derived from my previous work (Buschmann 2008) or originated based upon the theoretical propositions developed above (see chapter 1.2), with mixed results in light of the collected data. For example, my assumption that a viable and prosperous village life characterized by high rates of collective action and relatively high incomes – as identifed in detail by the CSS survey topic regarding (A) social and economic characteristics of each village – would lead to higher registration rates than among deprived villagers situated in loose networks turned out to be wrong. Further, I assumed there would be a relationship between respondents' primary source of information and their choice to register or not. Respondents whose information is primarily based on the internet (M4) would seem more likely to have registered their land, for information on land sales and further transactions referring to a plot of land are easily accessed online on the public registry's website (World Bank 2014). But this supposition likewise turned out to be wrong. These and further presumptions were later tested within the framework of simple descriptive statistics gathered through the survey (see chapter 5.1).

The variables were at first presented according to their overall frequency distribution to gain an understanding of the characteristics of the actors involved. To test whether correlations exists between the identified socio-economic factors and the rate of registration, each variable's frequency distribution was then calculated separately, according to whether land was registered or not. Both figures (registered or not registered) are then entered into an Excel sheet and transformed into

---

Qodalo, Kharadjala, Kardanakhi, Gavaza, Shilda, Kabali, Qesalo, Aghakla, Birliki, Lemshveniera, Tamarisi, Gurifhuli, Khulishkari, and Akhalsofeli.

[39] Ibid. (personal communication from, March 23, 2012).

graphs for ease in comparing both results. In cases where a correlation became obvious, the respective null hypothesis was tested according to its level of measurement.[42] If distinctive features resulted in calculations that gave reason for surprise, or that could be further enriched by information obtained during my post-survey field work in Georgia in 2013 and 2014, the analysis was accompanied by more detailed calculations.

2.2.3 Limitations

Overall, it needs to be underlined that, even if the total number of 600 (N) CSS' survey respondents may give reason to believe representative results could have been generated, the subsequent number of respondents, especially when formed into sub-groups of the sample population (n), was generally rather small. Moreover, looking carefully through the CSS (2012) data set, it becomes obvious that many more respondents were included who were located in regions in the East, (i.e. Shida Kartli, Kvemo Kartli and Kakheti), than those in the West (i.e. Guria, Imereti and Samegrelo).[44] For example, only one village in Samegrelo – Khulishkari – is taken into consideration, while data on the other two, Akhalsopeli and Guripuli, is missing. Thus, seeing that Georgian regions were neither evenly taken into account nor displayed proportionally to their size in terms of number of respondents, it does not seem that the survey produced a comprehensive picture of rural areas throughout Georgia. Moreover, the use of stratified sampling to select a sample population may have biased the study's results, as respondents were not chosen randomly. With the explicit aim to include the village intelligentsia, the actual composition of village populations is not likely to have been accurately reflected, which may have influenced the results. The same also applies to groups, such as the less endowed, maybe even landless rural dwellers, who were not taken into consideration. Finally, seeing that examination of the survey results was done through simple bivariate analysis, based on frequency distributions, their generalizability regarding the behavior of participating actors and their land-registration choices is limited. Nonetheless, the survey results have served as primary indicators spotlighting actor properties and their associated transactions, which were studied more thoroughly via the qualitative research process explained below.

---

[42] The variables A1, B16, R5 and F19 were measured on an interval scale, and their null hypotheses were accordingly tested with a t-test; since the remaining variables were classified on a nominal scale, their null hypotheses were tested via a chi-square test (McHugh 2013: 143). The null hypothesis was rejected for results at a significance level of $p = <0.05$.

[44] CSS (2012) chose the following villages located in the East: Aghaiani and Vaka (Shida Kartli); Lemshvenieri, Tamarisi, Aghtaklia and Birliki (Kvemo Kartli) as well as Kabali, Kodalo, Kardenakhi and Shilda (Kakheti); the following villages were chosen located in the West: Shorisubani and Nasakirali (Guria); Kulishkari (Samegrelo); as well as Koreti and Kesalo (Imereti) (information obtained by email, March 23, 2012).

2.3 Qualitative fieldwork: Case study design

Analyzing the structure and resulting outcomes of diverse property arrangements can appear easy on paper but may turn out to be difficult in the field. Thus, "[a] carefully crafted case study is another method for analyzing more complex action situations and their linkages" (Ostrom 2005: 35).

A case study is defined "as an intensive study of a single unit for the purpose of understanding a larger class of (similar) units" (Gerring 2004: 342). A "unit" refers to "a spatially bounded phenomenon – e.g., a nation-state, revolution, political party, election, or person – observed at a single point in time or over some delimited period of time" (ibid. 2004: 342). An aggregate of (un)studied cases forms a "population", and a "case" "is comprised of several relevant dimensions ('variables'), each of which is built upon an 'observation' or 'observations'" (ibid.: 342).[46] The case study approach "enjoy[s] a natural advantage in research of an exploratory nature" (ibid.: 349), seeing that the most pronounced feature of the case study method is its analytical depth (ibid.: 348). Noor (2008: 1603) therefore indicates that the technique "enables the researcher to gain a holistic view of a certain phenomenon or series of events [...] and can provide a round picture since many sources of evidence were used". In addition, the method allows for "capturing the emergent and immanent properties of life in organizations and the ebb and flow of organizational activity, especially where it is changing very vast" (ibid. 2008: 1603).

2.3.1 Selection of agri-produces and study sites

In the preparatory phase for conducting a field study in 2013, I established contact with the International School for Economics (ISET), located in Tbilisi, who offered working space and integration into the school's upcoming events. It was here that the author attended a presentation by Jacques Fleury, vice-president of the Georgian Wine Association, CEO of Georgian Wines & Spirits Company (GWS), director of the wine brand Château Mukhrani, and former director of the mineral-water bottling company Borjomi, a Georgian brand known throughout the former Soviet Union. He outlined the development of the company after its privatization and his challenges in terms of pushing counterfeit products out of the market and reestablishing the brand on local as well as international markets. In a subsequent private discussion about my intended research in Georgia, Mr. Fleury invited me to attend negotiations between GWS and a local wine company, Golden Khvanchkara, regarding contested land borders for the titling of vineyards in Racha, a mountainous region in the North-East of Georgia. Starting with GWS

[46] It might, hence, be "helpful to conceptualize observations as cells, variables as columns, cases as rows, and units as either groups of cases or individual cases (depending upon the proposition and the analysis)" (Gerring 2004: 342).

and Golden Khvanchkara as a pilot study, the analysis proceeded on the basis of snowball sampling; two further wine companies were then interviewed upon suggestion, the first, Telavis Gvinis Marani, which was founded by Georgian partners, and Schuchmann, a German investor in Kakheti.

Subsequently, I established contacts in villages that had already been included in the survey undertaken by CSS the year before (cf. CSS 2012):[48] four villages (1Q, 2K, 3S, 4G) of Kakheti, a wine region in Easterm Georgia as well as three hazelnut producing villages located in the West, Samegrelo (5K, 6K, 70). These villages were chosen based upon my knowledge that the first two exhibited relatively high registration rates among respondents included in the CSS survey, whereas all of the other five villages had displayed low rates. With contact established, focus groups were organized among small-scale wine and hazelnut producers as well as semi-open interviews with wine producers and hazelnut processors in these villages. Moreover, I interviewed local, regional and central government representatives as well as experts and representatives from international organizations.

During a period of expert interviews in Georgia's capital, Tbilisi, I interviewed Deloitte Consultant Vincent Morabito and FAO's National Project Manager Lasha Dolidze, land-tenure experts for USAID, who both had established links to hazelnut producers and processors, among them the best-known being Ferrero, the largest hazelnut producer in Georgia. Moreover, help was subsequently provided by USAID's branch in Zugdidi, the capital of the western-located region of Samegrelo, which within the framework of a joint project with Ferrero, gives support to hazelnut producers and processors in the region (Edilashvili 2012). Thus, during the field trip to Samegrelo, based on the help of USAID's local project manager and two other staff members, I interviewed representatives of four companies engaged in hazelnut processing, conducted three focus groups with nine small-scale hazelnut producers and interviewed one member of the *gamgebeli,* head of the local administration in Samegrelo, as well as a member of the *sakrebulo*, the Municipal council of Zugdidi (*Zugdidis mucipalitetis sakrebulo*), a regional administrative state organ.

The qualitative data for my analysis was rounded out by expert and government-representative interviews with the following persons: the Deputy Head of Amelioration Department of the Ministry of Agriculture; an employee of the National Agency for Public Registry (NAPR); the Agriculture Attaché at EC Delegation in Georgia; an academic professional working in the field of land titling of the Division of Human Geography at Javakhishvili Tbilisi State University; two land-tenure experts from USAID; a former director of the State Department for Land Management (SDLM), who is now working at the German Gesellschaft für

[48] With the aim to "trying to maximise participant anonymity alongside maintaining the integrity of our data" (Saunders et al. 2015), the villages' names are disclosed; the original names can be obtained on request from the author.

Internationale Zusammenarbeit (GIZ); as well as two representatives from Transparency International Georgia (TI Georgia) and an expert of the Georgian Young Lawyer's Association (GYLA).

Overview of field visits: two field visits took place in 2013 and 2014, with a total of 46 interviews and focus groups. Among them, during the 2013 visit, the 16 interviews took place with experts and (local) government representatives. Seven focus groups were held with grape- or hazelnut-producing farmers in Kakheti and Samegrelo as well as seven semi-open interviews with wine producers and hazelnut processors (see table 2-1). In a follow-up visit in 2014, further interviews took place again with at least one member of the same focus groups, as well as further experts and government representatives.

Type of interview	2013	2014
expert interviews	11	5
government representatives	5	3
wine producers (East)	3	1
hazelnut producers (West)	4	1
focus groups (East)	3	4
focus groups (West)	4	4
total number of interviews	30	16
total length (in hrs)	29:08:07	11:32:05

Table 2-1: Interviews carried out in Georgia in 2013 and 2014

The interviews and focus groups conducted in 2013 and 2014 had the following specific aims:

Expert interviews: becoming familiar with the interviewees' personal backgrounds and perceptions – in terms of benefits and problems – of the early land distribution and its documentation, property registrations, supporting measures by the international donor community, and the rather newly-established cadastre system and its organizing entity NAPR (on central and local level), as well as their understanding of specific spatial or landtype-related characteristics affecting land tenure or governance.

Government representatives: understanding interviewees' positions (central/local/regional level), their perceptions of policies on the local level vis-á-vis the distribution, privatization, registration and governance of land, as well as their views on the effects of land registration on local agrarian production.

Interviews with wine producers and hazelnut processors: understanding the founding stories of their companies, links to the government, employment schemes (overview of human resources), buildings, inputs and land used; their acquisition of natural resources; and any problems related to the sector.

Focus groups: getting an overview of participating villages, their histories, agrarian products as well as agricultural production today; gaining familiarity with inhabitant's incomes and (off-farm) employment; obtaining information on the general marketing channels as well as specifically on intermediaries that farmers usually rely on; understanding agrarian production during the Soviet period as well as afterwards, during the process of restructuring and reorganization; getting an overview of inhabitant's land resources, property rights, land-registration rates, and the effects of land registration on agricultural production. Particular questions related to people's land registration – spanning from the early times when the land was allocated until today – were clarified in a second field visit, in interviews with at least one focus group representative (contact person) in 2014.

Study sites: based on the results of the first, quantitative analysis (see chapter 5.1, figure 5-2) on the local distribution of registration rates in Georgia, two villages with high and two with low registration rates were chosen in both regions, Kakheti (1Q, 2K, 3S, 4G) and Samegrelo (5K, 6K, 70).[49] My first field visit took place from June 18 to July 2, 2013, beginning in Tbilisi by interviewing experts, including representatives of international organizations (e.g. Transparency International Georgia and GIZ), academic personal from Ilya Chavchavadze University as well as practitioners (e.g. lawyers). Then, from June 21 to July 2, 2013, research was focused on the study site at Kakheti where, on the one hand, semi-open interviews took place with three different wine producers and local government representatives and, on the other hand, four focus groups were organized with five to ten farmers each. After further expert interviews were held in Tbilisi, research in the field continued from July 8 to 12, 2013 in Samegrelo, where semi-open interviews with representatives from the local government, the hazelnut-sector supporting organization USAID as well as four hazelnut processors – among them Ferrero – were held in or near Zugdidi, the region's administrative capital. Eventually, three focus groups with five to ten hazelnut-producing farmers each were also organized.

The second research stay took place from July 27 to August 6, 2014 and was used to clarify open questions and shed light on vague matters. Follow-up questions were, therefore, addressed to experts, government representatives and focus group members with whom I had previously met during the first visit.

2.3.2 Overview of case studies

As the overall object of the present study is to examine land privatization in Georgia, two agrarian sectors, namely wine and hazelnuts, were chosen for my case studies, because they not only represent two of the leading export sectors of Georgian agriculture (DWVG 2013b) but also heavily depend on securing of their property rights in the long run. Additionally, these products may serve as good examples for gaining representative results, for hazelnut cultivation is located in Samegrelo, Western Georgia, while wine production is concentrated in Kakheti,

[49] Due to time limitations and organizational challenges, only three villages in Samegrelo were finally included as study sites, in contrast to the four villages in Kakheti.

Eastern Georgia (DWVG 2013) (see figure 2-2 below).[51] Thus, choosing both commodities has led to the creation of a research design that takes into account spatial variation. Individual small-scale agricultural producers as well as (leading) processing companies in both fields were studied.[52] As already described above, during the Soviet era Georgia (*sakartvelo*) "used to be one of the largest producers of wines, apples, pears and citruses as well as tea leaves" (Bluashvili & Sukhanskaya 2015: 12). Today, the key export products are "hazelnuts (29%), wine (16%), mineral and fresh water (13%), alcoholic beverages (11%), cattle (3%), non-alcoholic sparkling beverages (3%), [and] live sheep (2%)" (MoA 2015: 82).

Figure 2-2: Regional map of Georgia (Znovs 2011)

[51] The map of Znovs (2011) provides a clear overview of Georgia's regions; the outlines of the Russian annexed areas may not precisely reflect those in reality.

[52] Although this study was intended to focus primarily on permanent crops, i.e. hazelnuts and wine grapes, during the course of the field research in July to August 2013 it turned out that the production of these crops was in practice not separated from production of other plants. Hence, my analysis includes the land tenure of arable and pasture land, too.

2.3.2.1 Case study I: Wine (Kakheti)

Due to its favorable climate, fertile soils and cheap labor makes Georgia a premier wine destination (Mamardashvili 2008: 12); for an illustration of a Georgian traditional bowl for wine storage (*qvevri*) of about 1900, see figure 2-3 below).

Figure 2-3: Qvevri – traditional bowls for wine storage, about 1900 (Georgian National Museum 2014)

Kakheti spans nine climate zones: a humid climate in the Alazani floodplain (and other lowlands of Lagodekhi), a temperate humid climate in dry subtropics (Shida Kakheti), temperate dry steppes (Gare Kakheti) spreading through most of the territory, and, finally, a zone of permanent snow and glaciers in the high mountain climate of the Greater Caucasus watershed (UNDP Georgia 2014: 7). Beside grapes, apples and tangerines constitute local traditional products that "are also among the top three permanent crops in terms of the volume of production" (Mamardashvili 2008: 20). But, "[w]ith more than 500 varieties of grapes grown in Georgia and one of the oldest wine making traditions, grape production is believed to have a potential to lead the growth in agricultural sector" (ibid. 2008: 21). Today, about 30 to 40 wine companies exist in Georgia, working collectively in the recently formed Georgian Wine Association.[54]

Georgian wine is exported to 41 countries, although the five major target countries (Russia, Kazakhstan, Ukraine, China and Poland) take a share of 86% of the approximately 24 million bottles sold annually (Kaufmann 2015). As already manifested during the Soviet era, Georgian wine exports heavily depend on the Russian and Ukrainian markets (which together add up to about 60% of sales), making this one-sided focus extremely vulnerable to external shocks, as the Russian ban on Georgian mineral water and agrarian produce (2006–2013) has shown (ibid. 2015: 2; see table 2-2 for an overview of Georgian grape production (1992–2013)).

[54] According to J. Fleury (personal communication, June 21, 2013).

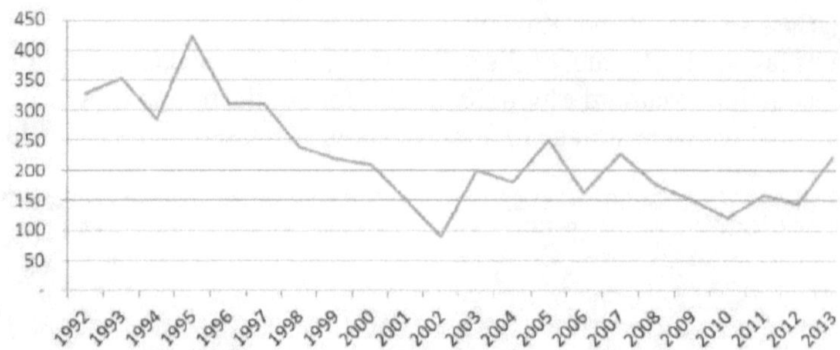

Table 2-2: Production volume of grapes in Georgia (1992–2013) (in thousand tonnes) (Bluashvili & Sukhanskaya 2015: 21)

The wine sector is subsidized annually, based on a fixed price in accordance with the quantity of grapes produced, not their quality. Until 2012, the government paid half of this fixed sales price to farmers; in 2013 and 2014, the subsidy was provided to wine producers; and since 2015 the policy has benefited farmers again, cutting the purchasing price of grapes for wine producers in half. Together with low sales figures on the Russian and Ukrainian markets (e.g. in 2015), the policy might eventually lead to overproduction of grapes and, hence, a price decline for wine (ibid.: 3). The situation is aggravated by the fact that from 2015 the government guaranteed farmers the purchase of all unsold grapes (for a lower price) to be processed in distilleries leased for this purpose, putting likewise downward pressure on the price for pure alcohol (ibid.). The government hands out compensations to farmers in case of lost harvests due to hail damages, incidents that mainly hit Georgia's Eastern agrarian areas (MoA 2015). Additional disadvantages of the local sector listed by Mamardashvili (2008: 2) are the sector's small-scale production, its poor infrastructure as well as the growers' lack of expertise and capital.

Located in Eastern Georgia, Kakheti is said to be "the country's primary wine region" (Burton 2016: 90): It produces more than half of the total grape production; Imereti and Shida Kartli are among the other main producer (Bluashvili & Sukhanskaya 2015: 39). The region – formerly belonging to the Kingdom of Iberia (*kartli*) and from 1762 the united Kingdom of Kartli-Kakheti alternately under Persian, Ottoman and Georgian suzerainty till its annexation in 1801 by the Russian Empire (Suny 1994: 80–81) – involves the mountainous province Tusheti in the North, and a historical province in the South that contains its administrative center, Telavi (Kaufmann 2003: 204; Mühlfried 2010). The area is comprised by two great river valleys, i.e. the Alazani River in the North (Inner Kakheti) and the Iori River basin (Outer Kakheti) in the South (ibid. 2003: 204).

The latter, becoming more steppe-like toward the South, is used for husbandry and wheat cultivation; the Northern area, divided by the Gombori mountain range, is characterized as a huge orchard and vineyard (ibid.: 204).

	2006	2007	2008	2009	2010	2011	2012	2013
Imereti	36.3	54.5	43.7	30.3	25	26.3	36.2	36.6
Shida Kartli	10.9	16	8.1	16.4	8.6	10.2	13.6	18.7
Kakheti	80.2	118.6	100	82.7	64.7	98.1	70.8	129.5
Other regions	35.1	38.2	24	20.7	22.4	25	23.3	38.1
Total	162.5	227.3	175.8	150.1	120.7	159.6	144	222.8

Table 2-3: Regional grape production in Georgia (2006–2013) (in thousand tonnes) (Bluashvili & Sukhanskaya 2015: 20)

As the regional breakdown of Georgian grape production shows (see table 2-3), the sector experienced from 2006 to 2013 strong fluctuations within the last decade, with its lowest overall production in 2010. Kakheti produces with about 129.000 tons by far most of the grapes in Georgia. About a third (36.600 tons) of Georgian production is produced in Imereti, followed by Shida Kartli (18.700 tons) and production in other regions (38.100 tons). In 2013, about a third (36,600 tons) of Georgian production was produced in Imereti, followed by Shida Kartli (18,700 tons) and production in other regions (38,100 tons).

Agrarian production in Kakheti depends, however, on irrigation infrastructure which was, according to Bezemer & Davis (2013: 7) destroyed after the collapse of the Soviet Union. At the time, the plundering of resources allegedly including the most lucrative economic strategy of still-functional machines being sold as scrap metal to Iran and Turkey (Christophe 2004b: 90). In 2015, Kakheti was home to 178 wine companies which, altogether, processed 100,000 tons of grapes (GWA 2016). Still, wine making "is a long, slow process. It can take a full three years to get from the initial planting of a brand-new grapevine through the first harvest, and the first vintage might not be bottled for another two years after that" (Apallas 2016).

> In Georgia, the grape harvest is called rtveli. It begins in autumn, by the end of September, and lasts for several weeks. (…) Due to 80 percent of Georgia's vineyards being located in Kakheti, the eastern part of the country, this is the place where rtveli begins. The name itself stems from ancient Georgian stveli, meaning "fruit harvest." Over time, "s" was replaced with "r" and the meaning was narrowed down to grapes, thus giving birth to "rtveli". (Kajrishvili 2014)

2.3.2.2 Case study II: Hazelnuts (Samegrelo)

Next to wine, hazelnuts represent Georgia's key export product due to the country's favorable natural conditions, making it "the fifth largest producer of hazelnuts in the world following Turkey, Italy, US and Azerbaijan" (Bluashvili & Sukhanskaya 2015: 39). As the numbers show (see table 2-4 below), local hazelnut production "increased at a compound annual growth rate (CAGR) of 6% over 2006 – 2014, from 24,000 to 37,000 tons" (Kurdadze 2015). As of 2015, about 50 local companies were purchasing hazelnuts from growers for export (ibid. 2015).

According to the FAO, the total world production of shelled hazelnuts is estimated in 2013 with 858,697 tons, "with Turkey having a dominant market share of 63.9%, followed by Italy (13.1%), USA (4.7%), Georgia (4.6%), and Azerbaijan (3.6%)".[57] In terms of total hazelnut planting area, in 2014 Georgia ranked third place after Turkey and Italy (Kurdadze 2015).

In addition to its favorable climatic and natural conditions making hazelnut production a traditional good, Georgia's competitive advantages also include its "unique diversity of endemic hazelnut varieties" (ibid. 2015: 1479), high quality, and reasonable quality–price ratio (ibid.: 1479). Around 10% of locally produced hazelnuts is consumed domestically, while "[m]ost of the hazelnut suppliers on Georgian market are small households, there are some large farmers as well, their market share is relatively small but steadily increasing" (ibid. 2015: 1480). In fact, prospects for Georgian hazelnut production are considered to be quite good:

> Georgian hazelnut growers are dependent on world market conditions, but in the long term, hazelnut demand is expected to increase and Georgia has a real shot at becoming one of the top 3 hazelnut suppliers in the world. Ferrero alone is a multi-billion dollar consumer of hazelnuts, with each jar of Nutella containing up to 50 hazelnuts (~170gr). (Kurdadze 2015)

The export of hazelnuts in 2013 added up "to 116.4 million USD (increase of 128% and 65.3 million USD)" (Chavleishvili 2015: 1481). In total,

> [r]oughly ¾ of Georgian hazelnut exports went to the EU, where prices were 43% higher on average than in the CIS countries (2014 World Bank estimates). Italy, Germany, and Spain jointly imported 46% of hazelnuts exported from Georgia. The largest importer from the CIS countries was Kazakhstan, accounting for 8% of Georgian hazelnut exports. (Kurdadze 2015)

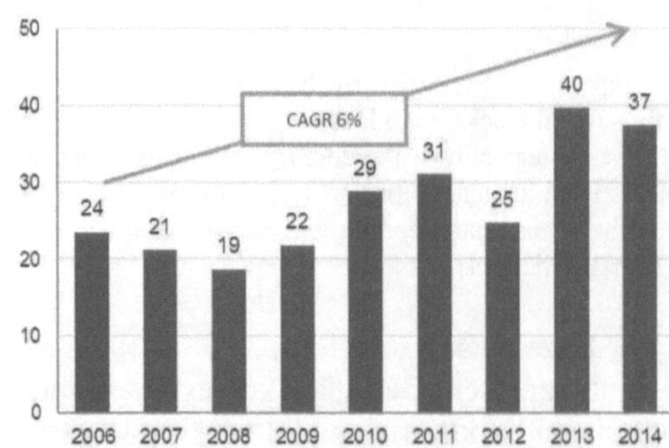

Table 2-4: Hazelnut production in Georgia (2006–2014) (Kurdadze 2015)

---

[57] It has to be emphasized that "Turkey is a price setter on the hazelnut market. In 2014 when the hazelnut harvest suffered from a severe frost in Turkey and output almost halved, hazelnut prices on the world market doubled from US$ 5.5 to US$ 11.5" (Kurdadze 2015).

Regionally, hazelnut production is centered in Western Georgia (see figure 2-4 below), with more than half (57%) of the country's total production in Samegrelo and Zemo Svaneti, whereas Guria, produces 20%, Imereti 11%, and other regions accounting for 12% (Kurdadze 2015).[58] As Chavleishvili explains,

> [h]azelnut production is one of the oldest sector[s] in Georgian agriculture. Hazelnut has been grown along the Black Sea and in the Caucasus since ancient times. According to historical sources, growing hazelnut in this region began in the 6th century BC. (Chavleishvili 2015: 1479)

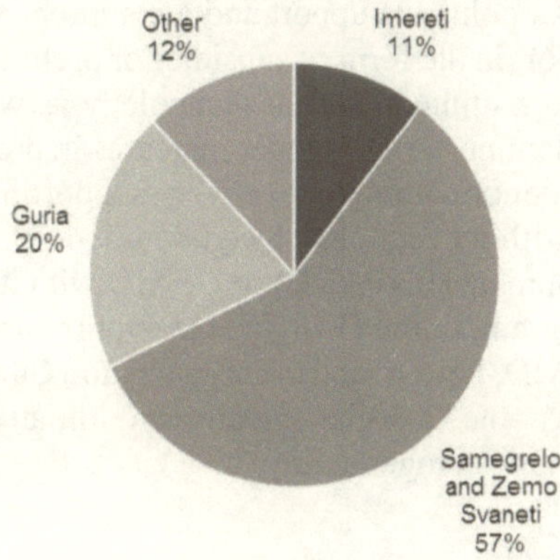

Figure 2-4: Regional hazelnut production in Georgia (Kurdadze 2015)

As with grapes for wine, hazelnut production also takes some time to set in motion, as "hazelnut orchards need up to 7 years to reach their full yield capacity" (ibid.). Nonetheless, in Georgia, the production of hazelnut is economically vital, for the "[e]xploitation period of hazelnut tree[s] is 25–40 years, it provides high yield for 20–25 years […]. Growing hazelnut doesn't require heavy investment and it's less labor-intensive…" (Chavleishvili 2015: 1480).

Most of Georgia's hazelnut processing companies are concentrated in Samegrelo (Chavleishvili 2015: 1482). The territory, located along the Black Sea, belonged to the ancient Kingdom of Colchis (*egrissi*) where, later, Ottoman Turks and the Persian Safavids fought for hegemony over the region, until Samegrelo was eventually incorporated into the Russian Empire in 1803 (Suny 1994: 47). The area is geographically divided by the Likhi Range (about 1000–2000 meters

[58] According to Chavleishvili (2015: 1480), local hazelnut varieties consist of "Gulshishvela, Shveliskura, Khachapura, Dedoplis Titi, Anaklian and others. Mose Janashvili, a 19th century famous Georgian historian, described four hazelnut varieties had been grown in Georgia since ancient times, these were: 1. Early Hazelnuts; 2. Georgian-large; 3. Greek hazelnut; 4. Wild Hazelnut […]. Hazelnut was exported from west Georgia to other countries; more land area was covered with hazelnut orchards than currently".

above sea level) in central Georgia; Eastern and Western Georgia belong to different climatic zones, with a subtropical environment in Samegrelo for flora and fauna (Kaufmann 2003: 276). In terms of its populace, "Samegrelo is mainly inhabited by Megrelians. They speak a dialect of Georgian known as Mingrelian and nearly all speak fluent Georgian. Mingrelians are one of the largest Georgian subgroups, making up 23% of the whole population" (GSH 2017).

In 2016, the Government of Georgia enacted technical regulations for hazelnut products to promote "strict quality control which includes not only laboratory tests, but storage, packaging, transportation control rules as well" (MoA 2016). But even though future prospects have been positively evaluated for the Georgian hazelnut sector, it lacks political support and a comprehensive state strategy, including financial support in the form of subsidies or preferential loans. Moreover, fragmented land plots, a limited level of technology as well as the need for research (centers) and plant nurseries hamper increases in production and sales. Beside the international donor community actively supporting the sector – notably, the European Union with its European Neighbourhood Programme for Agriculture and Rural Development (ENPARD) or USAID with its Rural Economic Development (RED) program, or the Georgia Hazelnut Improvement Project, a collaboration among USAID, Ferrero and the organization Cultivating New Frontiers in Agriculture (CNFA) –the Georgian government initiated support to the sector in 2015 with a drying and storage facility.

2.3.3 Limitations

The limitations of Grounded theory, also referred to as an art (*Kunstlehre*) is challenging with view to the fact that it is a highly creative process that can hardly be learned but must be trained in the course of the analysis and thus demands what Michael Polanyi (1967) has termed as tacit knowledge (see chapter 3.3.3). Moreover, the necessity, "which seems initially to be liberating – that one should distance oneself from existing theories and allow the theory to grow out of the data, often causes insecurity among students" (Böhm 2004: 274). Collaboration with other (more experienced) researchers is thus of vital importance. Another source of ambiguity is built in the case study approach which "concerns the blurry line between a unit that is intensively studied – the case study – and other adjacent units that may be brought into the analysis in a less structured manner" (Gerring 2004: 344). The author emphasizes "…that because a case study refers to a set of units broader than the one immediately under study, a writer must have some knowledge of these additional units… Case studies are not immaculately conceived; additional units always loom in the back-ground" (ibid. 2004: 344). The complexity of within-unit analysis may turn out to be particularly tricky as "within-unit cases are often multiple and ambiguous" (ibid.: 344). Thus, Gerring (2004) reminds that, on the one hand, "[i]n many instances, ambiguities can be removed simply by more careful attention to the task of specification" (Gerring

2001, 90–99). However, on the other hand, the researcher "...should be clear about which propositions are intended to describe the unit under study and which are intended to apply to a broader set of units" (Gerring 2004: 345). This so-called *structural ambiguity* rests on the fact that "the utility of the single-unit study rests partly on its double functions. One wishes to know both what is particular to that unit and what is general about it" (ibid.). However, it is this very "opportunity to study a single unit in great depth that constitutes one of the primary virtues of the case study method" (ibid.).

3 THEORIES ON CHANGING THE RULES OF THE GAME

The key to the choice of institutions lies in the way that they enable us to save transctions costs. Fundamentally, this is the economic rationale for their existence.
(Dahlman 1980: 138)

As explained in chapter 1.3, the present thesis is based upon an analysis of institutional change. The theory of institutions applied here is based on a theory of human behavior that has been brought together with a theory of the costs of exchange, also known as transaction costs (North 1990b: 27). The following section provides an overview of the analytical components and theoretical foundations I employ to empirically describe the institutional change of property arrangements regarding agricultural land in Georgia. In concluding the chapter, the Institutions of Sustainability (IoS) framework is introduced, as it offers analytical tools for shedding light on and evaluating the outcomes of land privatization in Georgia.

3.1 Institutions

Human behavior is determined by institutional arrangements, namely prevailing rule structures that "provide incentives and constraints for individual actions" (Hodgson 2006: 6). As such, institutions are understood as both objective structures "out there" as well as subjective springs of "the human head" (ibid. 2006: 8). Stress is thus placed on both agency and the institutional structure where it is understood that institutions are in itself the outcome of human interactions; while not being consciously defined in detail by individuals or groups, given institutions historically precede status-quo aspirations of individuals (ibid.). As such they comprise set of rules, which are exhibited in (1) norms and conventions, (2) working rules (of going concerns) and (3) property relations, or entitlements (Bromley 2006: 51–62). Common to all three kinds of institutions is their correlative character by providing choice sets for individual and group behavior that define what is socially acceptable (ibid. 2006: 51). The specificities are described as follows.

3.1.1 Norms and conventions

Informal rules, like *norms* and *conventions* are social institutions which bring about regularity and conformity in everyday's behavior of actors, by establishing "a structured set of expectations about behavior (…), driven by shared and dominant preferences for the ultimate outcome" (Bromley 1989: 42).[59] The concept is

[59] This statement is supported by Arrow (2000b: 12) who stresses that "our expectations of the future affect what we do in the present".

closely linked to "what it means to be socialized into a particular culture" (ibid. 2006: 51), seeing that an "inculcation of such behavioral expectations, and their enforcement, resides in the family and its logical extensions" (ibid. 2006: 51). As such, they result in a matching of behaviors that is directed toward common ends that "bring order, civility and predictability to human relationships" (ibid. 51).

3.1.2 Working rules

Working rules are based on the concept of *going concerns*, which comprise e.g. firms, (non-)governmental organizations, families or nation-states (Commons 1934). The principle of going concerns is "more or less control of individual action by collective action. This control of the acts of one individual always results in, and is intended to result in, a benefit to other individuals" (ibid.: 70). The formation of these rules is based on the "cumulative volitional creation of those who have consecutively possessed the power or delegated authority to decide about upon the content of the concern's working rules" (Bromley 2006: 26; Ramstad 1990: 87).[60] Thus, *working rules* are generated by custom or law and form the scaffolding provided e.g. by boards of directors or legislators who define opportunity-sets or fields of action for the members of a going concern (Bromley 2006: 31). As a result, working rules – in accordance to Commons (1934: 71) – indicate what "[i]ndividuals *must* or *must not* do (compulsion or duty), what they *may* do without interference from other individuals (privilege or liberty), [or] what they *can* do with the aid of collective power (capacity or right)" (Bromley & Yao 2007: 11). Hence, working rules, on the other hand, also define "what they *cannot* expect the collective power to do in their behalf (incapacity or liability)" (ibid. 2007: 11). A specific kind of institution are *organizations* that are broadly understood "as social systems with boundaries and rules" (Hodgson 2006: 9). According to North (1990b: 73), organizations are "designed by their creators to maximize wealth, income, or other objectives defined by the opportunities afforded by the institutional structure of the society". Bromley (1989b: 43) thus emphasizes that organizations' working rules convey legitimacy to the outside of the regime and "articulate the necessary steps which must be followed to become a corporation, and to remain one" (ibid. 1989b: 43), as is e.g. outlined in constitutions, legislation, a charter or by administrative rules. Moreover, working rules evenly convey coercion that is effective inside of the corporation, among its members (ibid.: 43). The second type e.g. assigns employees' terms of appointment or how decisions must be made, and hence mark internal rules. "The former institutions define the corporation vis-à-vis the larger society, the latter give it structure" (ibid.; italics in

---

[60] Collective action is broadly understood as a governance structure that is found in all kinds of hierarchical authority structures, "that parameterize (define and redefine) individual realms of choice" (Bromley 2006: 36) by law making entities to be found in villages, clans, or any other going concern (ibid. 2006: 36–37).

the original).[61] By pursuing their objectives organizations change an institutional structure incrementally (North 1990b: 73).[62] Emphasis is placed on the specific interactions between institutions and organizations (ibid. 1990: 5). A final distinction is made between *integrative* and *segregative* institutions, a concept related to nature-related transactions, which stresses the imposition of (social) costs on others (Hagedorn 2005). Specifically, in the present line of argument these concepts are better referred to as *integrative* or *segregative working rules*, seeing that these terms implicitly comprise rights of one party, and a correlated duty of another. Hagedorn (2005) specifies that, on the one hand, "[i]ntegrative institutions consist of [working] rules which hold decision makers liable for the transaction costs they cause, they have the duty to internalise them and no right to externalise them" (ibid. 2005: 6). Segregative institutions, on the other hand, "soften this restriction maybe to different degrees and relief decision makers from transaction costs and burden others with them" (ibid. 2005: 6).

3.1.3 Property relations (or entitlements)

The third type of institutional arrangements is represented by property relations (or entitlements) that are related to "income (or benefit) streams arising from the ownership of particular valuable objects or circumstances (…) – usually (but not only) associated with the ownership of land" (Bromley 2006: 54). Property relations are rooted in the Hohfeldian legal correlates right, duty, privilege and no-right (ibid. 2006: 54–55), that are linked to valuable objects or income streams. Blomley (ibid.: 55) thus stresses that "[t]o have a right with respect to a stream of future economic benefits is to have the capacity to compel the state to protect – and perhaps indemnify if necessary – your control over that income stream…". In other words, "[t]he essence – the empirical content – of ownership is the socially sanctioned ability to exclude others" (ibid.). Property relations are thus a triadic concept that includes (1) a person who is entitled to "own" a valuable object; (2) a valuable object; and (3) all others, which makes it a contentious social institution (ibid.).

Property on the one hand, may refer to a physical object, e.g. a house, or to a benefit stream, e.g. revenues of a (renewable) natural resource (Bromley 1989: 185). In the case of land, property pertains to land as an *asset* based on its renewable resource stock, as well as to its flow in the form of its *utility* (Nutzwert) that either capitalizes directly in the form of yields, or indirectly in the form of a land

[61] Eggertsson (2006: 18) lists the following criteria that define an organization in a more specific way: "(a) criteria to establish their boundaries and to distinguish their members from nonmembers, (b) principles of sovereignty concerning who is in charge, and (c) chains of command delineating responsibilities within the organization."

[62] It is assumed that not all organizational members act in the common good given that some members do act, while others not always act in the interest of the organization's set objectives but in their own interest (Hodgson 2006: 10).

rent (Loehr 2012: 838). To summarize, "[p]roperty (...) is a benefit (or income) stream, and a property right is a claim to a benefit stream that the state will agree to protect through the assignment of duty to others who may covet, or somehow interfere with, the benefit stream" (Bromley 1991: 2). The main feature of a right, however, is "its correlation with the notion of duty, its involvement with coercion, the fact that it may be concerned with either acts or omissions, and the fact that violations require restitution" (Becker 1980: 8). As a consequence, to enforce a right and the interest of those who hold a right against others, the state must "interfere": "The state must do something, for to do nothing is to side with the party protected by the status quo property arrangement" (Bromley 1991: 19; cited from (Samuels 1971, 1972).

Whereby ownership is a social concept which is based on "empirical evidence", meaning that the "the full force of government stands ready to protect my interest in the thing" (Bromley 1991: 21); *possession*, in contrast, is an "empirical phenomenon" in the sense that it is footed on "intuitive evidence", "because little extra work is required for an observer to make a plausible connection between the possession or use of the object (asset) and its belonging to the user" (ibid. 1990: 20).[63] The latter neither entails the backing of an authoritative regime, nor the recognition, and hence legitimacy of the larger community to acknowledge the belonging of the thing as is proclaimed (ibid.: 20–21). Bromley (1989: 205) thus remindes to draw a distinction "...between who has a right to use something, who has a right to the stream of benefits from something, and who controls the access to something". The author thus highlights the focus on the multiple dimensions of property relations as "essential to a clear understanding of property as a social institution" (ibid.: 205). On the other hand, "[o]wnership, the proof of which is known as 'title', is a fundamental legal concept" (Dale & McLaughlin 1999: 17). Holding the full bundle of all associated rights to the property (so-called *freehold* in Common Law jurisdictions) is the highest form of ownership (ibid.: 17).

A *property right*, in legal terms understood as a specific sort of right, "typically has something to do with the right to use, the right to transfer, and the right to exclude others from the thing owned" (Becker 1980: 18). Property rights thus present a bundle of rights, more precisely termed "proprietary rights" (ibid. 1980: 22). Honoré (1961) develops a concept that covers the full and most liberal notion of ownership, that emphasizes "that the owner is subject to characteristic prohibitions and limitations, and that ownership comprises at least one important incident independent of the owner's choice" (Honoré 1961: 371), namely a legal system that "provides some rules and procedures to attain these ends" (ibid. 1961: 361). The concept is based on the legal correlates developed by Hohfeld (1913; 1917) and emphasizes that "the concentration in the same person of the right (liberty) of using as one wishes, the right to exclude others, the power of alienating and an

[63] The recognition by others – as is present in Kant's concept of "intelligible possession" (in contrast to "sensible possession" as simple bodily appropriation) – is the determining criterion of property relations (Bromley 2006: 188–189; cf. Williams 1977).

immunity from expropriation is a cardinal feature of the institution" (Honoré 1961: 370–371). For the sake of simplicity, the following analysis focuses primarily on Eggertsson's (1990) aggregates, who distinguishes the above mentioned bundles of rights into three broad categories: the exclusive right to use an asset, the right to appropriate its economic value, as well as the right to sell or alienate that asset (Eggertsson 1990: 34–35). However, the distinction presented above focuses on the conditions of physical possession and its restrictions, which form choice sets of the holders of these rights. This view is thus expanded to clearly outline the foundations and, more importantly, the implications of property rights as policy instruments.

Consequently, transactions may now be defined as "the alienation and acquisition, between individuals, of the *rights* of future ownership of physical things, as determined by the collective working rules of society" (Commons 1934: 58; italics in the original). Institutions "define the status quo against which any collective action must be regarded" (Bromley 1989b: 38–39) – and hence set "the legal foundations of the economy" (ibid. 2006: ix):[64] "The economy as a set of ordered relations obtains its structure and operational character from these rules and conventions" (ibid. 1989: 78). The concept of public policy is accordingly concerned with two central aspects, (1) deciding about "institutional arrangements (entitlement structures) that both constrain and liberate individual action; and (2) searching for the boundary between autonomous (market-like) and collective decision making" (ibid.:34). The first denotes who may participate and who has to bear the cost; the second delimits alternative entitlement structures through determinants of market and non-market processes (Bromley 1989: 103). As a result, "the status quo structure of institutional arrangements is at the core of how collective choices are framed, and thus acted upon" (ibid.: 182). The features of and impacts from institutional arrangements thus define, first, the status-quo array of choices to each economic actor; second, the resulting relationship among economic actors; and, third, "…indicate who may do what to whom (…)" (ibid.: 49–50). Thus, in particular, the "…prevailing structure of norms, conventions, rules, practices, and laws […] shape or define the choice sets of individuals and groups in any economy" (ibid.), whereas the agglomeration of institutional arrangements define economic conditions at a specific point in time (ibid.).

The distribution of costs and benefits under an existing property arrangement may change dramatically with the appearance of new interest or technical opportunities (Bromley 2006: 57–58). Changing one institutional arrangement to another may result in the form of a new benefit stream emerging to one party and the imposition of unwanted costs to another, mostly referred to as externality or

[64] Common (1934: 72) stresses that "[c]ollective action is even more universal in the unorganized form of Custom that it is in the organized form of Concerns".

spill-over effect (ibid. 2006: 57–58).[65] An *externality*, in Pigou's (1920) terminology, is the divergence between private and social cost, where voluntary contractual arranements have been made in the form of market transactions, but "there still remain some interactions that ought to be internalized but which the market forces left to themselves cannot cope with" (Dahlman 1979: 141).

Seeing that, from the point of a wealth-maximizing agent, "the cost of carrying out the actual transaction is greater than the expected benefit" (ibid.: 141–142), two approaches are suggested to solve this kind of "market failure", i.e. either through some form of state intervention (this was referred to shortly in chapter 2.1.3), or "through a suitable establishment of appropriate markets" (ibid.: 141), where "economic agents can be made to take into account the side effects they generate" (ibid.: 141).[66] The "classical" remedy for these spillovers is linked to the first approach, a concept developed by (Pigou 1920) and his tax-subsidy that leaves the choice either "of imposing an excise tax on the output, or of offering an excise subsidy for reductions of output, of a good that generates an adverse spillover effect" (Mishan 1974; Pigou 1920). The latter necessitates not only the active role of the state to internalize these effects but, vis-à-vis joint costs, assumes physical proximity of the parties involved (Bromley 1989: 59). The other "traditional" mentioned option is advocated by Coase (1960) who stresses the reciprocal nature of external effects and their costs, and therefore suggests to establish a common ground for the parties involved "to bargain over who will be made to bear the joint costs" (ibid.).

Property rights, in North's terms, "are the rights individuals appropriate over their own labor and the goods and services they possess" (North 1990b: 33). The ability to appropriate resources depends on the nature of the institutional framework, namely "the legal rules, organizational forms, enforcement, and norms of behavior" (ibid. 1990: 33). However, with due consideration of the costs of exchange, rights stipulated in contractual arrangements "are never perfectly specified and enforced; some valued attributes are in the public domain and it pays individuals to devote resources to their capture" (ibid.: 33). For this reason, Demsetz (1967: 350) proposes that "property rights develop to internalize externalities when the gains of internalization become larger than the cost of internalization" (see chapter 3.2.1). However, this account ignores unwanted costs imposed on others, seeing that property arrangements define which costs "…might legally be ignored, and property legitimizes those costs that are so visited on others" (Bromley 1989: 212). As a result, it is at the heart of the matter who is causing such costs and on whom these costs fall: "*One person's government interference is another's government protection*" (ibid. 1989: 212; italics in the original).

---

[65] The concept of externalities is a central critique of market organization in the theories of welfare economics and economic policy in general (Buchanan & Stubblebine 1969).

[66] Bator (1958) distinguishes three forms of market failures, i.e. externalities, monopolies and public goods; see for further discussions Calabresi (1968).

The process for qualifying for ownership, e.g. when the state transfers exclusive ownership to private individuals, may "dissipate the potential rent from the land" (Eggertsson 1990: 93; Anderson & Hill 1983, 1975). This is because the costs arising with the assignment of property rights are one-time sunk cost and do not affect the stream of benefit; in contrast, the cost of enforcement vary, occur over time and are incurred by he state and individual right holders (Eggertsson 1990: 96). "High costs of enforcing rights may render exclusive ownership of a resource economically inviable" (ibid. 1990: 96). A reassignment of property rights comprises "additional costs to the state of assigning and maintaining the new structure of rights [that] must be taken into account" (ibid.: 108–109). Accordingly, when gains are anticipated from new outcomes, these gains must be related to the cost of institutional change (ibid.: 108). As a consequence, "even a relatively modest change in the structure of property rights, such as the reassignment of liability in a world of exclusive rights, affects both the productive capacity of the economy and the distribution of wealth…" (ibid.: 101). E.g. the ability to affect the income flow of an asset without being made liable for the full costs generated lowers the value of the very asset; on the contrary, an ownership structure "in which those parties who can influence the variability of particular attributes become residual claimants over those attributes" (North 1990b: 31) maximizes an asset's value: "In effect they are then responsible for their actions and have an incentive to maximize the potential gains from exchange" (ibid. 1990: 31). In this regard, Bromley (2006: 120) stresses that "public policy is nothing but a modification of the institutional structure of an economy that redefines choice sets (fields of action) for individuals". Moreover "[t]he legislature and the courts continually address the need for new institutional arrangements – driven by the recognition that the economy is always in the process of becoming" (ibid. 2006: ix).

The most proponent theory that account for the formalization of tenure is the Evolutionary Theory of Land Rights which is summarized by Platteau (1996). It presumes a change in relative factor prices that pushes the government to formalize tenure by introducing titling programs. It is based on the so-called property rights school associated with Coase (1960), Demsetz (1967), Cheung (1970) and Posner (1977) (see chapter 3.3.3).[67] For these authors, the decisive factor for choosing an economic system based on private property rights is scarcity which sets the subsequent need for exclusive, unambiguous ownership rights, henceto introducing a private property regime. In contrast, they argue, a property regime based on collective ownership raises incentives for overconsumption and mismanagement of resources, hence rent dissipation is to be expected. By arguing that an individual person that owns land "will attempt to maximize its present

[67] The following overview does not deal with the philosophic justifying conditions of private property; the latter though is usually based on arguments referring to first occupancy (Cicero, Rousseau, Kant); labor theory (Locke); traditional views (Aristotle, Hegel, Hume) or economic justifications (von Mises, Hayek, Posner) referring to utility, political liberty or to considerations of moral character (Becker 1980).

value" (Demsetz 1967: 355), "privately owned resources will always tend to be allocated to the highest value uses" (Furubotn & Pejovich 1972: 1140). Private property rights so serve as the most efficient property regime (Platteau 1996: 31).[68] However, the reasoning and effects of formalizing tenure in valuable assets emerges from a set of assumptions and implicit predictions (Bromley 2008b: 20). The formalization of property registrations in official records is understood as "...the issuance of titles to individuals (or families) now holding (possessing) housing and other land-based assets in an allegedly tenuous and quite insecure state" (Bromley 2009: 20). Moreover, it is highlighted that the "claimed insecurity of tenure is blamed for stifling investment in the assets now possessed rather than owned" (de Soto, 2000) (ibid. 2009: 20). Hence, this insecure possession of assets is supposed to find its remedy by the issuance of titles. The prediction is that it is the security through titles that "are claimed to allow individuals to gain access to official sources of credit – banks, credit unions, lending societies" (Bromley 2008b: 20). These outcomes – start a new business, improvements in housing or investment in agricultural production – are assumed means to approximate the goal of eradicating poverty. What is assumed by *security* vis-à-vis to using the land as collateral "concern[s] possession of rights to *transfer* land rather than security as such..." (Sjaastad & Bromley 2000: 372). Security is thus primarily associated with the *assurance* of rights (ibid. 2000: 372). Thus, insecurity is understood as the individual perception about the probabilities of violation and detection, while the "cost of insecurity finds expression through the effects of the behavioural modifications to which this perception gives rise" (ibid.: 372). A higher perception of risk "results in a higher discount rate for future returns, thereby reducing the net return to investment and investment volume. A higher discount rate due to uncertainty over ownership biases investment towards short-term projects" (Barrows & Roth 1989: 2). Consequently, for the purpose to encourage development the "security of tenure need not amount to ownership, nor need it last for all time" (Simpson 1976: 8). E.g. a tenant may be granted leasehold for an adequate amount of time to secure investments. Thus, security is given by "a period long enough to serve the purpose for which the land is to be used" (ibid. 1976: 8). The important point is that "[i]n those countries where individual property rights are recognized and the rule of law prevails, the courts will uphold occupation against anybody – including the State or the Government – other than a person who can prove a better right" (ibid.: 9). A title, or the formalization of ownership, therefore cannot be viewed as panacea to spur investments and eradicate poverty.

[68] A further discussion on property rights regime is discussed below (see chapter 3.2.1). The meaning of "efficient" or "efficiency" shall not further be discussed here, except from the very notions that, firstly, "efficiency, however defined, is dependent upon the institutional structure that gives meaning to costs and benefits, and that determines the incidence of those costs and benefits" (Bromley 1989: 32); besides, efficiency cannot be understood in mere economic terms, "but must firmly be put in the socio-historical context within which any particular institution is designed" (Dahlman 1980: 9).

3.1.4 Implications of entitlements

As was presented above, social arrangements, i.e. going concerns, are regulated by working rules that are either generated by custom or law (see chapter 3.1.2). In the first case, private orderings may be set-up in form of a contract, in the latter case by the legal system. Regarding the first, private arrangements are of many forms, ranging from installing fences to the following of social norms. With view to the latter, an individual's "sphere of decision amounts to the collection of all the possibilities he or she has to decide regarding the use of assets of whatever kind they may be" (Werin 2003: 5). These possibilities are generally understood as *rights*, or *property rights* which "are supported (more or less effectively) by law" (ibid. 2003: 5), while "almost all interaction among people relies on formal law to some extent and is influenced by it" (ibid.: 6); the opposite is similarly true, "law is framed under the influence of the private motives, wishes and inducements behind interaction among people" (ibid.). Accordingly, "law is not created in a vacuum. Much or most of it has evolved in response to the needs of economic life. There is a mutual interdependence, a "symbiotic" relation. Hence, there is a single, interrelated *economic-legal* system" (ibid.; italics in the original).

Land, in particular, has "two special characteristics which distinguish it from all other commodities known to commerce" (Simpson 1976: 5), namely it is *immovable* and *everlasting*: The first specific of immovability results in the fact that "it cannot be transferred from one person to the other; nor it can be possessed in the same way as something that can be actually handled and moved about" (ibid. 1976: 5). Thus, if my land is not accessible without crossing somebody's else land, it is my right to trespass it, a so-called *easement* (ibid.: 6). The land's other feature, to be everlasting, has two important consequences: On the one hand, rooted in Anglo-Saxon law, land does not only comprise the surface but the air above it and the soil below (ibid.: 5) E.g. land which was registered in England about a century ago in the so-called Domesday survey is "still the same land today; the individual proprietary units into which it was divided may have changed completely but today's parcels are made up of the same land…" (ibid.: 6). These lands hence have undergone a so-called *mutation*, i.e. boundary changes. Consequently, even if an owner has absolute ownership, i.e. enjoys the full bundle of rights over a piece of land – *fee simple* in English law (see below) terms – it "is often withheld or restricted, for public policy demands that land shall not be allowed to fall into the wrong hands (though what hands are wrong is, of course, capable of widely differing interpretation)" (ibid.: 8). In such a case of presumably public interest, it is the State's right of so-called *eminent domain* to exercise power of *compulsory acquisition*. In reality absolute ownership of land does hence not exist. However, "[i]t is this capacity of land to carry future interests (…), which has led to many of the involutions of land law" (Simpson 1976: 6).

The nature of legal relations among individuals is then predicated on the concept of working rules and further based on the work of (Hohfeld 1913, 1917). A *legal relation* is defined as a "societal recognition of a specific set of ordered relations among individuals" (Bromley 1991: 16). The total of all legal relations is specified according to Hohfeld (1913; 1917) by eight legal conceptual terms which line up into a group of entitlements, i.e. rights, privileges, powers and immunities, and their legal opposites, i.e. no-rights, duties, disabilities and liabilities (Harbison 1992). Accordingly, these legal relations are formed by their legal correlatives right-duty, privilege-no-right, power-liability and immunity-disability (Hohfeld 1917: 710; see table 3-1).

Jural Opposites	right / no-right	privilege / duty	power / disability	immunity / liability
Jural Correlatives	right / duty	privilege / no-right	power / liability	immunity / disability

Table 3-1: The general nature of legal relations (Hohfeld 1917: 710)

A right is defined as an expectation on the side of the right-holder, who rest assured that others do not adversely affect it (Bromley 1991: 16); a right is associated with the duty to respect that right (ibid. 16–17). A privilege describes a situation where the right-holder is free to act toward the right-regarder; being confronted with a privileged counterpart leaves others with no-right (ibid.: 17). Having the power to act means to be able to establish a new legal relation toward others; the latter become subject to the new rule and, moreover, are made liable to that supremacy (ibid.). Enjoying immunity, on the other hand, means that, whatever legal relation is set up, one is not affected by it; being faced with the immunity of others means having a disability (or no power) to force those to act (ibid.).69 These correlates can be classified into two further categories, being either active (positive) or passive (negative) legal relations (Hoebel 1942: 956): The first two correlates, right-duty and privilege-no-right, are so-called active vis-à-vis the state authorities, seeing that "they are imperative relations subject to the coercive authority of the courts and other recognized law-enforcing agencies" (ibid.). "Hohfeldian claims [rights], privileges, duties and no-claims [no-rights] are "the *substance* of that insubstantial thing, the law"" (Harbison 1992; Llewellyn 1960, 1960). The latter two correlates, i.e. power-liability and immunity-disability, are *passive* in the sense that "[t]hey are not in themselves subject to direct legal enforcement. (…) Rather, they set the limits of the law's activities" (ibid.), where "the law itself declares to be outside the scope of its sphere of control" (ibid.). In other words, one has "to recognize that passive legal relations are statements of "no-law"." (Hoebel 1942: 956). However, along with the first, salient feature of a right's relational character, another key point is emphasized (Harbison

[69] With regard to the fact that individuals mostly belong to several sub-groups in society, these legal relations might overlap within an individual's choice set (Hoebel 1942: 956).

1992: 461): Claims and immunity are, secondly, historically contingent but, nonetheless, "the property rights the government will enforce at any given time" (ibid.).[70]

A right in the strictest sense is thus to have a right to call for coercive measures in case of denial of a sound claim; here, one possesses a *claim-right* toward others, whereas the latter have the duty to act accordingly, or to offer "compensation in lieu of it" (Becker 1980: 11).[71] A special sort of claim-right is a *capacity-claim* "to call particular attention to one's status as a potential *holder* of rights" (ibid.: 12; italics in the original), e.g. expressed by saying 'to have a right to get married' (ibid.). This is due to the fact that having a right often presupposes a needed *capacity* or *standing* to do something, which hence assumes other sorts of rights to exist (ibid.). The issue shall be made clear by the example of a football player who has a right to score (ibid.): On the one hand, it is based on a person's liberty (privilege) to score, since no one is allowed (no right) to stop somebody scoring; and it is based on power in the sense that if a player manages to maneuver the ball in the net in the agreed way, the goal has to be acknowledged by others (liability). Thus, a right based on liberty "entails only the absence of claim-rights in others" (ibid.: 13; italics in the original); a right based on power "entails only liability in others" (ibid.).[72] Thus, to give (legal) emphasis to one's liberty or power by securing that no one interferes with the one's liberty or power, the one might become protected by a claim-right, and a duty in others may be enforced (ibid.).[73] A final cause of confusion refers to *recipient rights*, a kind of right "which has some of the stringency of a claim right (the right-holder is 'owed' or 'entitled' to something) but for which no corresponding duty-bearers can be specified" (ibid.: 14). An example is to assume the right to health care, being rather a sort of moral claim

70 With Hohfeld's terms as point of departure, Harbison (1992: 461) outlines that concepts as easement or license "do not convey the full meaning of either social relations or legal rights", seeing that "they are not derivable from apriori abstractions (…) and (…) are at best convenient pigeonholes where legal analysis begins rather than ends. At worst, they are inconvenient" (ibid. 1992: 461–462).

71 The author offers further classifications that are not treated here (see Becker 1980: 11–12).

72 Liberties are distinguished between *natural* and derived, *institutional* liberties (ibid. 1980: 13): The first refer to those that "exist independently of any social institutions" (ibid.: 13), mostly pronounced among those "in the proverbial state of nature" (ibid.), where "no one has any claim-rights against them for the performance or nonperformance of any act. Once political or social institutions arise, natural liberties are limited (…), but political and social liberties of various sorts become possible" (ibid.). The latter hence depends "on the existence of political and social institutions" (ibid.).

73 As already mentioned, if my right is based (purely) on power, then I have the ability "to alter my relations to others (with respect to rights)" (ibid.). A more profound distinction is made between perfect and imperfect power (see ibid.: 13–14): The first relates to one's sole ability to alter a legal relation vis-à-vis others; the latter refers to situations which in fact demand the involvement of others, and is thus termed more precisely *participant-powers*.

though (ibid.: 15). Disputes over natural resources, being it fishermen who compete for diminishing fishing grounds, a plant which pumps its chemical residues into a river that formerly served as fresh water source or conflicting views on land use – widely perceived as "environmental problems" –, are to be understood as "problems of conflicting right claims" (Bromley 1991: 3). A factory that perceives to "have a right" to dump its chemicals into a river is an example of *presumptive* rights that are confronted "not to have been a right at all but merely a practice (ibid. 1989: 213). Based on Hohfeldian terms, the intruder so far enjoys a privilege, whereas the claimant has no right.

The denial of a right claim, especially with reference to natural resources or environmental issues, is thus either based on a lack of existence or coverage (ibid.: 4): Disagreements over the *existence* of a right denotes situations where two or more parties dispute over a specific kind of right, e.g. rights to preserve a beautiful landscape, or rights for quality controls of air and water; disagreements over a right's *coverage* concerns questions whether one claimant is indeed entitled to seek for protection by state authorities, e.g. my neighbor's right claim to enlarge her house versus mine to keep a pleasant view without disturbances.

Thus, by drawing closer attention to the entitlement structure based on the legal correlates developed by Hohfeld (1913; 1917) it is possible to characterize the underlying legal relationship, and hence to specify "[t]he nature of interdependence and interference among economic agents" (Bromley 1989: 208). The concept of entitlements (see chapter 3.1.3) is therefore extended by entitlements protected by property rules, liability rules and inalienability rules (Calabresi & Melamed 1972). Depending on the fact who enjoys protection, who interferes and the kind of protection provided (by property, liability or inalienability rules), result in different legal structures that bring about severe implications.

Property rules consist of right-duty correlatives that permit protection to the right holder. If one wishes to interfere with that right, the other not only has to initiate the bargain (ex ante) and bear all transaction costs incurred, but needs to offer a reservation price to convince the right holder to abandon the right. The inference may occur if the parties agree on a price to pay. It is hence a collective decision who is initially entitled to enjoy protection, but the bargaining parties decide on the amount to pay (Calabresi & Melamed 1972: 1092).

Liability rules comprise the correlatives privilege-no right, that permit others to interfere, and require compensation to stop them; the one without a (claim) right has to bear the cost of interference, initiate the bargain (ex post), cover all transaction cost incurred and pay a price to compensate the privileged one to stop the action. The value required to stop the inference is "set by the neutral eye of a third party (usually the state)" (Bromley 1989: 210); its payment depends on the one's willingness to pay.

Inalienability rules are defined by the legal correlates immunity-no power, whose aim is to protect unwanted third party effects and hence may preclude a transaction. On the other hand, inalienability rules may become relevant (apart from third-party effects) if "the action is characterized by irreversibilities, and if we are not certain of our benefit measures for such things into the future" (Bromley 1978: 48). At this point, "the inalienability rule may be preferred" (ibid. 1978: 48). In this case, the state intervenes to determine the initial entitlements as well as "the compensation that must be paid if the entitlement is taken or destroyed, but also to forbid its sale under some or all circumstances" (Calabresi & Melamed 1972: 1092–1093). Thus, unlike property or liability rules, this rule provide protection to an entitlement, but its granting is regulated and limited by the state (ibid. 1972: 1093). As is summarized by Demsetz (1967: 347), "property rights specify how persons may be benefited and harmed, and, therefore, who must pay whom to modify the actions taken by persons". The question arises under which conditions these alternative entitlements, to wit property or liability rules, should be applied? E.g. protecting the interest of many individuals, who need to bargain prior an offending action under a property rule, seems rather impossible; on the contrary, ex post bargaining is more practical under a liability rule if the offending party may proceed but is obliged to pay compensation after the action (Bromley 1989: 213). The latter moreover shows that the tendency to interfere is boosted in the case of liability rules, as it is easier to proceed (the courts decide on compensation payments thereafter), than to bargain for a price to obtain prior consent (ibid.: 213–214).

The way an entitlement is protected thus affects (a) the time of the bargain to take place (ex ante or ex post), decides on (b) who has to initiate the bargain and (c) who has to bear the costs of both the inference and the transaction (villain versus claimant), (d) who is involved in litigation (the relevant parties versus state organs) and depending on this aggregate, (e) how probable an inference might be (see table 3-2).

	property rule		liability rule		inalienability rule	
	right	duty	power	no right	immunity	no power
(a) bargaining	-	ex ante	-	ex post	-	-
(b) initiated by whom	-	x	-	x		
(c) transaction costs	-	x	-	x	-	x
(d) litigation costs	-	x	-	x	-	x
(e) tendency of inference	low		high		low	

Table 3-2: Effects of property, liability and inalienability rules (adapted by Bromley 1989: 206–216; 1991: 42–51)

The overview shows that the initial assignment of rights is decisive vis-à-vis a person's legal status and the ensuing distribution of costs:[74] A claimant is better

[74] The initial endowment of resources relates to property rights and the individual's wealth position. A critique on Coase (1960) exactly focuses on his assumption that an alternative initial assignment of rights has no wealth effects if volitional bargaining is possible, for it

protected against unwanted cost under property rules, which pass the costs of inference on the intruder before an action may take place; the expenses are covered privately by the villain. The state authorities assign the initial property arrangements, but no further state intervention is required. Liability rules, in contrast, ease disturbances by allowing preceding inferences to take place, while its' settlement is passed onto state organs and the costs incurred are covered collectively. Finally, inalienability does not provide for any disturbances; it is a cost-saving measure for the benefit of the public and the individual, seeing that both are strictly freed from unwanted costs. Property relations are hence contentious seeing that they carry "the ability to hold something off the market until a possible buyer meets the price that the owner is free to set" (ibid. 1991: 37): Any attempt to change the entitlement structure either depends on the reservation price (property rules) or the willingness to pay (liability rules). Thus, seeing that it is income and current endowments that are the decisive factors for an entitlement structure to change (Bromley 1991: 48), it must be concluded, that the initial assignment of rights matter, and more precisely, "who it is that obtains the initial assignment of entitlements, and the nature of that entitlement" (ibid.: 37).

The imposition of rules which change institutional arrangements is, first, a conscious and determined measure in the framework of economic or public policy; likewise, property rights are policy instruments in the sense that by altering these rights among individuals or groups is to accomplish certain desired ends (Bromley 1991: 35). Second, it is necessary to keep in mind that existing entitlements "are simply artifacts of previous scarcities and priorities, and of the location of influence in the political process" (ibid.: 201). Thus, an important issue of economic analysis is to understand that changes in social values and priorities are in accordance to shifts in property rights (ibid.: 199). For an understanding of agricultural settings, it is thus necessary to keep in mind that transaction costs stem from a particular nature of the production processes; by showing the exact link between the productive technology and transaction costs it is possible to specify the nature of transaction costs "before we can account for the choice of property rights in any particular context" (Dahlman 1980: 138–139).

It could be learned that the status quo structure of institutional arrangements, which defines the choice domain of individuals and groups, is based on norms and conventions, the working rules of a going concern and is, when linked to income or benefit streams, specified by property relations (see chapter 2.1.4). Thus, by altering rights relationships among individuals or groups is to accomplish certain desired ends (Bromley 1991: 35). It follows then that "to analyze a proposed policy is to attempt to understand who the gainers and losers are, how they regard

is supposed that "the rights will go to the highest bidder and efficiency will be assured regardless" (Bromley 1991: 36). However, this view does not take the emergence of transaction cost into account, which e.g. lead to "imperfect, costly, and asymmetric information among participants to a bargain" (ibid. 1991: 37), and thus "will influence the ultimate distribution of welfare across members of society" (ibid.: 36).

their new situation *in their own terms*, and what it means for the full array of beneficial and harmful effects" (Bromley 1991: 219; italics in the original). However, what needs to be carefully kept in mind when analyzing individuals' choice sets is what prospect theory has shown, namely that individuals "have a greater distaste for losses from a status quo position than for the gains that may arise from changes in that status quo" (ibid. 2006: 74; (Kahneman & Tversky 1979; Tversky & Kahneman 1974). The ensuing difficulty here is to distinguish between both an individual's preference and an individual's choice: A *preference* is viewed as a state of mind vis-à-vis an outcome or a prospect, whereas a *choice* relates "to an act, a decision, or a strategy" (Bromley 1989: 85). Preferences and choices are hence dependent on an individual's choice which is defined by the status quo: "The concern here is to explore the relationship between preferences and choices, and those situations in which goals or objectives dominate preferences, and those in which preferences are dominated instead by means or instruments (ibid. 1989: 85). Thus, Shubik (1982) points to the challenge to make out the distinction between preferences among *outcomes* and preferences among *strategies* (ibid.). As a result, the scientific task is to address policy impacts relevant to those that are affected and, specifically, "how individuals regard the benefits and costs of certain policy *alternatives*" (Bromley 1991: 220). With view to the fact that the main idea behind collective action and institutional change is a process that transforms certain interest into entitlements, it is essential to distinguish between (desired) policy objectives and policy instruments chosen for this goal (ibid. 1991: 227).[75] "This distinction presumes that decision makers first choose policy objectives, and only then begin to search for policy instruments to achieve those objectives" (ibid.).[76] Whether formalization of property rights in land is viewed as a specific policy tool or as policy objective is treated more carefully in the next parts.

3.2 Governance

3.2.1 Governance of nature-related transactions

Governance structures are "organisational solutions" for institutionalized transactions (Hagedorn 2008: 360) that "enforce rights and the corresponding duties of others to respect those rights" (Di Gregorio et al. 2008: 6). *Property regimes*, i.e. private, state or common property regimes (see figure 3-1 below), are instrumental human artifacts, i.e. collective perceptions in terms of defining "what is scarce

[75] By referring to transform *interest* into entitlements indicates that an individual (or a group) "has some strong feeling about a particular situation; they have a *stake* in the situation at hand" (Bromley 1991: 227; italics in the original).

[76] It must be highlighted though that it is not always easy to draw a sharp line between both policy instruments and policy objectives; on the contrary, in reality "decision makers often will start with existing activities and gradually define and formulate objectives in view of experience with policies" (ibid. 1991: 227; cf. Blaug (1980: 151).

(and hence *possibly* worth protecting with rights), and what is valuable (and hence *certainly* worth protecting with rights)" (Bromley 1991: 3; italics in the original).[77] Regimes are understood as social constructs "whose purpose is to *manage people* in their use of environmental resources" (ibid. 1991: 21; italics in the original). A property regime proves to be successful if "(1) the natural resource has not been squandered; (2) some level of investment in the natural resource has occurred; and (3) the co-owners of the resource are not in a perpetual state of anarchy" (ibid.: 21). With view to the contested nature of governing resource use, the role of environmental policy is understood as to challenge a "putative rights structure that gives protection to mutually exclusive uses of certain environmental resources" (ibid.).

State property
Individuals have the duty to observe use/access rules determined by the controlling/managing agency. Agencies have the right to determine use/access rules.

Private property
Individuals have the right to undertake socially acceptable uses, and have the duty to refrain from socially unacceptable uses. Others (called "non- owners") have the duty to refrain from preventing socially acceptable uses, and have a right to expect only socially acceptable uses will occur.

Common property
The management group (the "owners") has the right to exclude non- members, and nonmembers have the duty to abide by exclusion. Individual members of the management group (the "co-owners") have both rights and duties with respect to use rates and maintenance of the thing owned.

Non-property
No defined group of users or "owners", the benefit stream is available to anyone. Individuals have both privilege and no right with respect to use rates and maintenance of the asset. The asset is an "open-access resource."

Figure 3-1: Four types of property regimes (adapted from Bromley 1989a: 872)

Depending on the property regime individuals enjoy rights and duties in a different way: *Private property* provides rights to individuals to use and have access to a benefit stream and implies others having a duty to respect that right; *common property* regimes provide a limited number of individuals with rights and duties to use and maintaining a benefit stream according to rules established by its members; non-members have the duty to respect to being excluded; under a regime of *state-property* individuals have the duty to recognize the rules to a benefit stream

[77] Though scarcity is viewed as "the central, controlling fact of the contexts in which problems about property rights arise" (Becker 1980: 6), the author gives rise to concern that "like many seemingly obvious general pronouncements, it is not so clear, upon inspection, that this one is as uncontestable as it seems" (ibid. 1980: 6). It may be, he stresses, that "one's possession has some importance just because those things are one's own appropriations – regardless of whether or not they are scarce or likely to become scarce" (ibid.: 6).

set by a controlling agency, i.e. an authorative state body; *open-access* resources are termed those reserves whose access is not limited and whose use is not controlled; individuals possess both no right and no privilege vis-á-vis accessing or using the resource (Bromley 1989a: 872).

The social construction of institutions and their associated governance system depend "on the property of the transactions and the characteristics of the actors involved in such transactions" (Hagedorn 2008: 359; Schmid 2004: 69ff.). These so-called institutional arrangements hence "emerge either spontaneously through self-organisation or intentionally by human design" (ibid.). Regularized behavior is mediated by norms and conventions and thus evolves endogenously through individuals' interaction (Bromley & Yao 2007: 11).[78] In particular, it is about "...those institutions at the "informal" end of the spectrum – the norms, habits, standard practices, customs, traditions, and conventions that provide important boundaries to, and parameters for, much individual and group action..." (Bromley 2006: 22–23). In contrast, working rules and entitlement regimes "...are policy variables and therefore subject to change through collective action (Bromley 2005: 4). In other words, they are "imposed institutional arrangements [that] are the conscious and purposeful instruments of national (and provincial) economic policy" (ibid. 2007: 11).

With due consideration of transactions – that "contain in itself the three principles of conflict, mutuality, and order" (Hagedorn 2008: 363; cf. Commons 1932) – the interdependence of social and natural systems may either result in friction or coherence that both produce certain cost: The first relate to *transaction costs* which are typically viewed as an "economic equivalent of friction in physical systems" (Williamson 1985: 19). These are the costs to gain information, the costs to contract with one or the other, and the costs to enforce contracts (Bromley 1991: 105).[79] Besides, Bromley (1991) highlights that transaction costs "are much higher in sparsely settled rural areas" (ibid. 1991: 105) than in urban neighborhoods, seeing that the former are typically characterized by institutional preconditions which guide individual activities with "strategic uncertainty" that does not allow for "providing a secure basis of economic calculation over space and time" (ibid.: 105). In contrast, the "rural hinterland" is viewed by the latter as "either irrelevant or a domain to be plundered for the benefit of the urban classes from whence power and legitimacy flow. (...) Destruction of natural resources is the logical outcome of these circumstances, and such practices continue because they serve the interest of the national government" (ibid.). This phenomenon is also known as "urban bias". The second, *coherence costs*, typically occur in social systems with "coordination and consensus building between many stakeholders"

---

[78] Among the three different classes of institutions it is entitlement regimes that are – even though subject to change by collective action – considered as the most explicit and durable institutional relations (Bromley 2006: 180ff.).

(Hagedorn 2008: 362). In any case, Bromley (1991) reminds that "the very notion of "transaction costs" is culturally specific – one person's tedious meeting (a cost) may be another's most enjoyable activity (a benefit)" (Bromley 1991: 31). Consequently, "there is no presumption that the invisible hand will result in harmony over the rules to be followed, instead it is the court system that must decide disputes and *create* order, or a "workable mutuality," out of conflict" (Commons 1934: xxx; italics in the original). Accordingly, "governance is an effort to craft *order*, thereby to mitigate *conflict* and realise mutual *gains*" (ibid. 1932: 4; italics in the original).[80] Or, in the words of Williamson (1996), "...the study of incomplete contracting in its entirety implicates both ex ante incentive alignment and ex post administration (which is what governance is all about)" (Williamson 1996).

Thus, to make an institutional choice among the many types of property regimes needs to take into consideration "the costs of the particular property regime pertinent for each type of land" (Bromley 1989a: 869). As will be shown below (see chapter 3.2.3), the formalization of ownership rights to agricultural land is thus not only about registering owners and their interest to landed property, but needs to ascertaining existing rights pertaining to the land plot together with a proof of evidence, while at the same time, set up and maintenance cost occur for the demarcation, indication and surveyance of the ground in question. In contrast, the administration cost for setting and holding up a collective property regime comprise meetings among villagers to agree on the resource's "...specific locations of use, to discuss rates of harvesting, and so forth" (Bromley 1989: 869). As a consequence, "[t]here is a reasonable case to be made that the costs of privatization (fences, measurement, title insurance, record keeping) are greater than those of the collectively-managed public domain of the village (ibid.), especially in environments where people, all knowing each other, are dependent on using public domaine land. The latter point, however, does not exclude that some sort of use rights exists to the resource stock or its flow; a combination of commonly used and privately owned strips of land is known, for instance, from open field systems (see e.g. Dahlman 1980), where e.g. the scattering of (privately owned) land "constituted the least costly way" (Bromley 1989b:17): Scattering granted the retention of "the collective decision making necessary to realize the returns to scale in livestock" (Dahlman 1980: 129). This is because "[s]cattering achieves a change in constraints and incentives, rather than creating negative rewards in terms of punishment for keeping out of the collective" (ibid. 1980: 129). Again, the "multiple dimension of the property relation is essential to a clear understanding of property as a social construction" (Bromley 1989b: 205).

[80] "Order" in Commons (1934: 6) sense refers to 'Futurity' and its anticipation, and hence inhibits a "security of expectations" which is "based upon reliable inferences drawn from experiences of the past; and also from the fact that it may properly be said that man lives in the future but acts in the present" (ibid.: 57–58).

3.2.2 Polycentric systems

The alignment of rules in accordance with particular properties of (ecosystem-related) transactions has brought a diversity of intuitional arrangements (Hagedorn 2008: 363): In classic transaction cost economics organizational forms range from buy-or-make decisions, thus via markets ("buy") or within firms by vertically integrated organization, so-called hierarchies ("make") (Williamson 1981). The decision is dependent on the required mechanisms for coordination and control, as well as the ability to adapt to disturbances (Williamson 1991). Accordingly, markets and hierarchies rather underlie the marketing of private goods and commodities within the above-mentioned state-market dichotomy (Hagedorn 2008: 363–366). In contrast, many examples concerned with governing nature-related transactions (which allow for interdependency among actors) show that tailored governance structures, hybrids, are required "because neither the command and control mechanism ["make"] nor the price mechanism ["buy"] can fulfil this task" (ibid.: 370), and hence emerge as hybrid, or polycentric governance system. The latter was firstly termed "polycentric" in the realms of public policy by V. Ostrom et al. (1961) and "connotes many centers of decision-making which are formally independent of each other" (ibid. 1961: 831). In the further discourse these alternative forms of governance were labeled as hybrids (Ménard 2004b).[81] Though hybrids do not necessarily share a common basis of ownership, they are comprised of "a great diversity of agreements among legally autonomous entities doing business together, mutually adjusting with little help from the price system, and sharing or exchanging technologies, capital, products, and services…" (ibid.: 348). Ostrom et al. (1961) specify three aspects "which give rise to public rather than private provision of certain goods and services" (Ostrom et al. 1961):[82] First, to control for externalities, i.e. spill-over effects that are not internalized privately need to be regulated by a public organization (ibid.; see chapter 2.2.2); second, as a consequence of decomposability, those who do not pay for a good or service may still enjoy its benefits (ibid.: 833). Both spill-over effects as well as decomposability result in others' non-exclusion of its (direct and indirect) effects, and thus mark a "jointness of consumption" and the absence of rivalry (Ostrom 1975). As it is outlined by Dewey (1927), "the public consists of all those who are affected by the indirect consequences of transactions to such an extent that it is deemed necessary to have those consequences systematically provided for" (Ostrom et al. 1961: 833). Third, the maintenance of a good or service as a preferred state of affaires focuses directly on the impact of the good's provisioning. "The exclusion principle provides a criterion for distinguishing most public goods

[81] See Menard (2004) on a thorough examination on (the emergence of) different forms of hybrids; according to his provocative words, a "collection of weirdos" (ibid. 2004: 347).

[82] Ostrom et al. (1961: 832) follow Dewey's (1927) definition that "…the line between private and public is to be drawn on the basis of the extent and scope of the consequences of acts which are so important as to need control whether by inhibition or by promotion".

from private, but it does not (…) clarify or specify the conditions which determine the patterns of organization in the public service economy" (ibid. 1961: 833).

Consequently, as a "preferred state of affaires" (ibid.: 833), declared as being in the interest of the public, it must specify the conditions which allow for the good's measurement and quantification: "The modification consists in extending the exclusion principle from an individual consumer to all the inhabitants of an area within designated boundaries" (ibid.). The calculation of benefits and cost of the provisioning is viewed as "a balance between the demands for public services and the complaints from taxpayers" (ibid.: 834). Thus, the *question of scale* arises with regard to the constitution and organization of its provision, namely how to specify the "boundaries of a local unit of government as the "package" in which public goods are provided" (ibid.: 835). In general, a public good is appropriately packaged (within its boundaries) when the exclusion of others to enjoy the public good is warranted (Ostrom et al. 1961: 833). In addition, the authors identify four criteria to designate the boundaries of a local unit, these are (1) *control*, (2) *efficiency*, (3) *political representation*, (4) *equitable distribution of costs and benefits*, and (5) *self-determination* (ibid.: 835):

1) *Control* refers to boundary conditions which "include the relevant set of events to be controlled" (ibid.); in case that proceedings are not matched within a fixed margin, the boundaries are adjusted to meet the criterion of control (ibid.).
2) The modification of a boundary meets the criteria of *efficiency* just as much as the appropriate technology is in use and the labor force is endowed with skills and proficiency to support the organization's undertaking (ibid.: 836).
3) The boundaries are further determined (in line with the underlying political interest) by means of *political representation* which affect the organization's performance directly and indirectly (ibid.): Political representation is, first, characterized by its *scale*, indicating the size and the feasible number of the governmental units that provide the public good, and "which specifies the formal boundaries" (ibid.); second, those affected by the public good comprise the *public*; third, the functioning of the organization is determined by a *political community* who actually decides "whether and how" (ibid.) the public good is provided. But, with view to the fact that "[t]hose who are affected by such a decision may be different from those who influence its making" (ibid.) the criteria of responsibility and accountability should be in line with democractic principles. Thus, "[w]here in fact the boundary conditions differ, scale problems arise" (ibid.): Depending on the transaction taken into consideration the *public* may differ while, vice-versa, the scale of the organization might be adapted to specific transactions and beneficiaries. "Similarly, a public organization may also be able to constitute political communities within its boundaries to deal with problems which affect only a subset of

the population" (ibid.).[83] Thus, informal arrangements may be introduced to modify the boundary conditions, so that "public organizations may (1) reconstitute themselves, (2) voluntarily cooperate, or, failing cooperation, (3) turn to other levels of government in a quest for an appropriate fit among the interests affecting and affected by the public transactions" (ibid.: 837). The authors underline that "[i]t would be a mistake to conclude that public organizations are of an inappropriate size until the informal mechanisms, which might permit larger or smaller political communities, are investigated" (ibid.: 836).

4) If direct and indirect beneficiaries of the organization's sphere of influence can be determined, the expenses for the good's provision are assessed to ensure warranting the *criterion of equitable distribution of costs and benefits* (ibid.).[84] Consequently, the authors highlight that "[e]xcept where a re-distribution of income is sought as a matter of public policy, an efficient allocation of economic resources is assured by he capacity to charge the costs of providing public goods and services to the beneficiaries" (ibid.).

5) Finally, the *criterion of self-determination* is decisive in determining the other four, namely control, efficient functioning, political representation and the equitable distribution of costs and benefits, seeing that "their application in any political system depends upon the particular institutions empowered to decide questions of scale" (ibid. 837).

Hence, home rule may demand a petition of the local citizenry or the approval of local government officials to address and remedy problems in the local domain; however, self-determination assumes that public goods are locally (successful) internalized, and do not become a matter for another political community (ibid.); if internalization is not possible, "and where control consequently, cannot be maintained, the local unit of government becomes another "interest" group in quest of public goods or potential public goods that spill over upon others beyond its borders (ibid.). Hence, in case that "internalisation is not complete, discrepancies between private and social costs and benefits will emerge (Pigou 1920)" (Sjaastad & Bromley 2000: 365). The set-up of municipal organization may thus be diverse for providing public goods and services among different self-governmental entities, and distinct arrangements "of local autonomy and home rule constitute substantial commitments to a polycentric system" (ibid.).

---

[83] With view to any operationalization, the author find a "one-to-one mapping of the formal public organization, the public and the political community as impractical" Ostrom et al. (1961), seeing that the public, depending on the transaction, changes .

[84] Ostrom et al. (1960) consider this criterion as belonging to the criterion of political representation; here, however, it shall be treated as an additional aspect for the actual analysis.

By hypothesizing on the nature of a polycentric system (and implicitly confronting it with "Gargantua", a so-called centralized political system) the following normative features are assumed (ibid.: 837–838):[85] (i) *Control*: Simplified bureaucratic and administrative structures are apt to respond to local (public) interests. (ii) *Efficiency*: The cost of maintaining control over boundary conditions, the appropriate technology and people's skills is minimized and secures the efficient provision of public goods. (iii) *Political representation* is the result of (a) a feasible number of, and reasonable nature of local government units (scale). The proper scale of local government units assures responsiveness to local public needs as well as to sub-sets of public interests by apt communication channels to remedy the "problem of "field" or "area" organization" (ibid.: 838). (b) Representation is assured among those who are affected (public). Accordingly, the organization further allows for the timely provision of the good as well as information thereon (allowing the formation of public opinion). (c) Representation finally provides for equal access of decision-makers to pursue in the common interest, resulting in unbiased policies so that "[m]unicipal reform may [not] become simply a matter of "throwing the rascals out"" (ibid. 837). (iv) The *equitable distribution of costs and benefits* is ensured by assessing the costs of public control and charge them upon the beneficiaries. (v) *Self-determination*: The institutional arrangements based on home rule and the degree of local autonomy (determining the question of scale and hence allow for control, the efficient functioning of the organization, political representation and the equitable distribution of costs and benefits), lead to an internalization of the public good.

The conditions to understand and evaluate the performance of a polycentric governance system are described by patterns of (1.) *cooperation*, (2.) *competition* and (3.) *conflict* among the interdependent parties involved in the governance system (Ostrom et al. 1961). 1.*Cooperation* refers, on the one hand, to the realization of joint activities that generate greater benefits to the actors concerned; and it relates, on the other, to the precondition that the "appropriate set of public interests are adequately represented" (ibid.). With reference to cooperation in hybrid organizations and the *pooling* of resources, Ménard (2004a: 354) raises the following concerns: First, to pool resources might give rise to opportunistic behavior among the partners; besides, decomposing tasks and endowments presupposes, second, joint planning and monitoring; third, the processing of information becomes a crucial issue among partners, and asymmetric information may provoke conflict.[86] He therefore stresses the implementation of control mechanisms (ibid.

85 The term "Gargantua" was firstly introduced by Wood (1958) as "the invention of a single metropolitan government or at least the establishment of a regional superstructure which points in that direction" (Ostrom et al. 1961: 831); it was taken over by Ostrom et al. to refer to urban infrastructure, as "harbor and airport facilities, mass transit, sanitary facilities and imported water supplies" (ibid. 1961: 837).

86 Ménard (2004: 352) points to continuity of a relationship as crucial condition to pool resources, and highlights that "continuity requires cooperation and coordination: partners

2004: 354–355), and highlights that "[h]ybrids operate as a buffer, with risk sharing as a central motivation" (ibid.: 355). 2. *Competition* in hybrid organizational arrangements "may produce substantial benefits by inducing self-regulating tendencies with pressure for more efficient solution in the operation of the whole system" (Ostrom et al. 1961: 838). With view to the fact that hybrids are composed of different providers of goods and services, polycentric organization "...may give rise to a quasi-market choice" (ibid. 1961: 838) among consumers demanding a good or service.[87] Ménard (2004) raises an objection to the advantage of rivalry within an organization by taking into account that the actors involved "remain independent residual claimants with full capacity to make autonomous decisions as a last resort" (Ménard 2004b: 352), which may provoke opportunistic behavior that is not in line with the interest of the whole organization (ibid. 2004: 352). Consequently, seeing the competitive nature of contractual arrangements within hybrids, Ostrom et al. (1961: 838) stress the potential of separating the *production* from the *provision* of goods and services. This might open up "the greatest possibility of redefining economic functions in a public service economy" (ibid. 1961: 838), for control is warranted by public provisioning in accordance with performance criteria, while competition may exist among those producing the good. The advantage is that "[b]y separating the production from the provision of public goods it may be possible to differentiate, unitize and measure the production while continuing to provide undifferentiated public goods to the citizen-consumer" (ibid.: 839). Especially when isolating the production from the provisioning of the public good, V. Ostrom (1975) underlines that "...alternative mechanisms also need to be created for articulating and aggregating preferences into a collective choice about the quality and/or quantity of the public good or service to satisfy the demands of the community of users" (Ostrom 1975: 691).

Ménard (2004) introduces the possibility of *contracting* as a formal mechanism to regulate partnerships (ibid. 2004: 352), which has the advantage of sharing competencies and scarce resources, but runs risk of uncertainties due to incomplete contracts and represents "a typical transaction-cost problem" (ibid.), where costly renegotiation is required. In addition, the "[r]eliance upon outside vendors to produce public services may reduce the degree of local political control exercised" (Ostrom et al. 1961: 840), which may "restrict quality and quantity of information about community affaires" (ibid.). As a result, cooperation presumes careful scrutiny, control, and regulation.

---

must accept losing part of the autonomy they would have under a market relationship without the benefits of extended control that hierarchy could provide".

87 However, to benefit from a situation of rivalry among equal providers assumes, on the one hand, that different providers of goods and services exist and, on the other hand, that actors get informed on the quality and prices rendered by others, according to the author's claim that "a number of units are located in close proximity to each other and where information about each other's performance is publicly available" (Ostrom et al. 1961).

3. *Conflict resolution* becomes necessary when the provision of a public good is not fixed within local boundaries and may lead to spill-over effects that might become a source of conflict among interdependent actors (ibid. 1961: 840). Accordingly, joint action is eased if benefits and costs are equally distributed throughout the area, and problems are "treated as uniform risk" (ibid. 1961: 840); if not equally shared, state regulations are required to internalize nuisances by sanctions or other enforcement mechanisms (ibid.). "The courts have thereby become the primary authorities for resolving conflicts…; and their decisions have come to provide many of the basic policies" (ibid.: 841), which "minimizes the risks of external control by a superior decision-maker" (ibid.). In contrast, the "[a]ppeal to central authorities runs the risk of placing greater control over local (…) affaires in agencies such as the state legislature, while at the same time reducing the capability of local governments for dealing with their problems in the local context" (ibid.: 842). Hence, [r]econciling the interdependence of actors (…) with their legal autonomy (…) is a key challenge in governing hybrid arrangements (Hagedorn 2008: 370).

3.2.3 Land administration and registration of ownership

The FAO defines *land administration* as "the way in which the rules of land tenure are applied and made operational" (FAO 2007).[88] More specifically, land administration "refers to the processes of recording and disseminating information about the ownership, value, and use of land and its associated resources (Dale & McLaren 1999: 859). In particular, these records include the specification (or adjudication) "of rights and other attributes of the land, the survey and description of these, their detailed documentation, and the provision of relevant information in support of land markets" (ibid. 1999: 859). In particular, adjudication is defined as a dispute resolution process, whereas registration defines the course of "making and keeping records of property rights" (Dale & McLaughlin 1999: 10). Accordingly, "Land Administration Systems [LAS] are the basis for conceptualizing rights, restrictions and responsibilities related to people, policies and places" (Enemark 2009: 1). The elements involved in land administration are illustrated by the following figure showing the "three key attributes of land" (Dale & McLaughlin 1999: 9), namely *land ownership, land values* and *land use*, to describe "the regulatory support system between the legal and policy basis of land, and the land (or real estate) market in which people participate" ((Dale & Baldwin 1998); see figure 3-2 below).[89]

[88] The term "land governance" used here refers to the organizational mode of how land and its use is regulated and administered; it is typically employed to classify (dichotomous) modes of governing land as 'weak' versus 'good' (see e.g. FAO 2007); "land administration" – as an established field of research – is used in the following to refer to the specific institutional arrangements of governing land.

[89] The following description is complemented by "The four basic components of land administration" by Dale & McLaughlin (1999: 8–12).

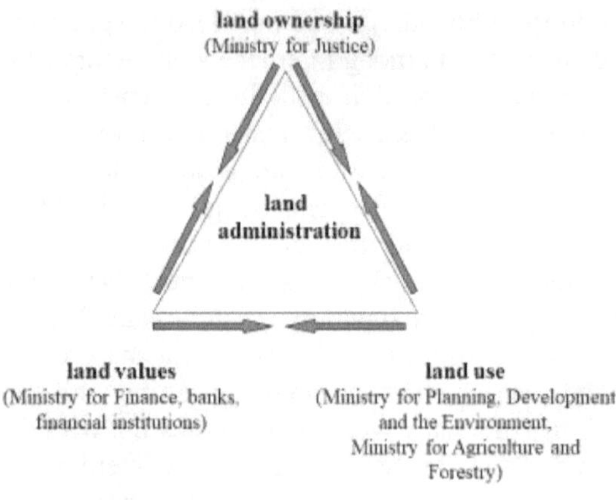

Figure 3-2: Three key attributes of land (adapted by Dale & McLaughlin 1999: 9)

The first attribute, *land ownership*, deals with land registration and is the juridical component that refers to people's rights and title; it mostly constitutes "a central government responsibility" (Dale & McLaughlin 1999: 8) and is typically located at the Ministry of Justice (ibid. 1999: 9). The second attribute, *land value*, is market-based and "a process that aims to establish the connection between monetary value and the property itself by producing an estimate of the capital value of the asset" (Dale & Baldwin 2000); it is the fiscal element and thus situated in the Ministry of Finance, whereas banks and other financial institutions may be evenly involved when being "responsible for revenue collection" (Dale & McLaughlin 1999: 8). *Land use*, the third attribute, refers to physical planning and is a regulatory component affecting the use of land. These elements are interdependent though seeing that "ownership affects the use of the land while conversely the use will influence the form and substance of the tenure. However, "[o]nly when all three pillars are in place within an economy can large-scale land registration be undertaken effectively, and can the formal property system rally sufficient social support (Cashin & McGrath 2006: 632). Seeing that land privatization describes an institutional change of property rights regimes from state to private ownership, the following section provides a more thorough picture on the first-mentioned juridical part, *ownership*, and its subsequent effects on land users' rights and title to finally studying its impact on agricultural production. *Land registration* is shortly defined as "the official, systematic process of managing information about land tenure" (Nichols 1993: 4).[90] For the purpose of this thesis, *land tenure* is defined

[90] As is outlined by Nichols (1993: 4), "a system is an organized set of components and relationships among those components. (…)", and highlights that it is designed "to meet objectives either internal to the system or defined by the external system environment. In a land registration system, for instance, to facilitate conveyancing is an external objective while to store information efficiently is an internal objective". Information, in this regard, "is data

as "*an institution encompassing the rights, responsibilities, and restraints that govern the allocation, use, and enjoyment of land*" (Nichols 1993: 7; italics in the original).[91] It is thus "a broad set of rules, some of which are formally defined through laws concerning property while others are determined by custom" (Dale & McLaughlin 1999). The *information* to be recorded relate to

(i) *people*, i.e. individuals or groups with a recognized interest in land;
(ii) *the nature of the interest* in land, "i.e. rights, responsibilities, and restrictions in land, including their duration and their effect" (Nichols 1993: 7);
(iii) *the land,* "i.e., the units of land, or land parcels, to which these interests apply, including location, value, resources, and use where appropriate" (ibid. 1993: 7).

The information is finally recorded "in a textual or graphical format and the medium may vary" (Nichols 1993: 7). The crux of the matter is that, first, land registrations are kept officially for resolving disputes or being used as collateral; and, the process of recording is realized systematically, i.e. "policies, standards, and procedures [are] in place to collect, validate, maintain, and provide access to the information" (ibid. 1993: 7). The latter, so-called *unit of record*, varies according to "[t]he purpose for which land registration is required (…), for it should determine the choice of the unit of record. If the purpose is fiscal, value may be the principle objective and the most suitable unit of record may be the *unit of use"* (Simpson 1976: 4), e.g. individual fields when it relates to agriculture, "which vary in size and quality and so in value" (ibid. 1976: 4). On the other hand, the units of use may be summarized in a *unit of operation*, that can make up a farm, "where development is concerned or the implementation of laws regulating land use" (ibid.: 5). Moreover, "[t]wo or more units of operation – whether contiguous or not – may, however, be comprised in a *unit of ownership*" (ibid.; italics in the original). The latter is mainly used "to give particulars of ownership and not of value or use" (ibid.). As is emphasized by Simpson (1976: 7) "the unit of record is clearly a crucial factor in land registration. Definition of the parcel and identification of those holding rights in it are the twin problems which dominate the subject". The challenge refers to possible subordinate interests to ownership that are not recorded, as e.g. an easement or other use-rights.

The aim of registration is "… to compile a complete register of all land, public as well as private, showing the ownership of every parcel and any limitation or restriction to which that ownership may be subject" (ibid.). The registration process thus demands safeguards, seeing that the compilation of existing rights may vary from place to place, or from country to country; land titles might originate from various sources, e.g. by an official grant or by mere occupation (ibid.: 188).

---

that has been processed and conveys a meaning or significance to the user with respect to a specific activity" (ibid. 1993: 5).

[91] The definition stems from "[u]npublished lecture notes in cadastral studies, Department of Surveying Engineering, UNB, Fredericton, N.B." (Nichols 1993: 23).

Consequently, Simpson (1976) reminds that before registering ownership it must be ensured that, first, the rights to be recorded are unambiguously defined; that, second, the seller is the original owner with the right to sell the land in question; and third, the buyer is informed on any subordinate or derivative interests by others in that piece of land (ibid.: 12–13).[92] Thus, it is about the proof of ownership, the so-called *proof of title* which moves into the focus of interest (ibid.: 14). Closely linked to the way in which transactions are confirmed and recorded is the legal status of the (previous) land registers, as the next illustration shows (see table 3-3 below).

Means of transaction		Evidence
oral agreement	--	witnesses
private conveyance	--	deed (no registration)
deeds registration	--	registration (no guarantee)
title registration	--	registration (proof of title)

Table 3-3: Types of transaction evidence (adapted by Larsson 1991: 17)

The registration of land is concerned with the *conveyancing* of land, i.e. the process to create and transfer interest in land (ibid.: 3). This method of transferring land is either based on *oral agreement* or *private conveyancing*, i.e. "conveyancing without recourse to any public records" (ibid.: 13). These forms of transfer were possible at times "when communities were small and close-knit" (ibid.: 13). The transfer in the presence of witnesses was mostly accompanied by a symbolic act, e.g. "handing over of a turf or a twig" (ibid.) which served as adequate evidence not only to the two parties, the seller and the buyer, but also to the public. With view to the latter, "many early systems of law (…) have regarded publicity alone as sufficiently effective guarantee when land is sold" (ibid.: 13). With the registration of a deed, it is assumed that when "society becomes more complex" (ibid.), symbolic acts as such do no longer spread among neighborhoods while third parties do no longer get knowledge on land dealings. Thus, land transfers are privately negotiated and then fixed by an officially written document (deed), that gives evidence to that the buyer "acquired the land from somebody who, in his turn, by production of the relevant document showed that he had acquired it from somebody who similarly proved his acquisition, and so on as far back as is required either by law or by custom" (ibid.13–14). The same process applies if the owner becomes an owner by succession, i.e. if the former ownership is proven sufficiently by a deed. A further step is the so-called *registration of deeds,* as proof of ownership. It is a kind of public register where "documents affecting interests

[92] The author points to *compulsory registration* of title as the only solution "to device a system under which conveyances of land can be conducted with the facility of sale of goods" (Simpson 1976: 13). A comparison between *compulsory* versus *voluntary registration* is treated further below.

in land are copied or abstracted" (Simpson 1976: 14–15). "Its basic principle in its simplest form is that registered deeds take priority over unregistered deeds, or deeds registered subsequently" (ibid. 1976: 15). The proof of ownership though "becomes a difficult technical process" (ibid.: 14), seeing that subordinate interests in land are not revealed among those documents dealing with purchases of the land in question, and "[e]ven the definition of the land itself is much more open to dispute when written description of the boundary is substituted for publicly walking around it" (ibid. 1967: 14). It hence demands "skilled investigation and practitioners learned in this special branch of the law" (ibid.: 14). Furthermore, private conveyancing has limitations, seeing that "above all it is not conclusive… In particular, there is the danger that, because all dealing has been secret, something which affects the title will not be discovered" (ibid.). As a result, a deed registration can be made a legal requirement "by providing in the law that unregistered deeds may not be received or admitted in court as evidence of title. Documents which are not registered can then be safely ignored, for they have no effect…" (ibid.: 15). Nevertheless, this process lacks legal viability seeing that "[a] deed does not in itself prove title; it is merely a record of an isolated transaction" (ibid.): It does not show whether both parties were vested with rights to this dealing, and it hence follows that an "investigation of its validity and effect will still be necessary before any further transaction can be safely conducted on the strength of it" (ibid.). Moreover, "a deeds registration will not show matters which affect a title but are not the subject of a deed. An example is succession on death, which gives title by operation of the law and not by act of the parties" (ibid.).

The so-called *registration of title* thus represents a "system which remedies the defects of registration of deeds; it enables title to be ascertained as a fact "instead of leaving it to be wrought out as an inference"" (ibid.). The register of title is defined as "… an authoritative record, kept in a public office, of the rights to clearly defined units of land as vested for the time being in some particular person or body, and of the limitations, if any, to which these rights are subject" (ibid.: 15–16). As such it gives legal validity to and about title, by providing the three safeguards mentioned above, namely, first, "unambiguous definition of the parcel of land effected (and any right over other land which is enjoyed in virtue of owning the parcel)" (ibid.: 16); second, contact information of the owner, namely an owner's name and address (or those information of a cooperation if it is the recognized owner); as well as, third, "particulars of any interest affecting the parcel, which is enjoyed by someone other than the owner" (ibid.: 17).[93] Sir Charles Fortescue-Brickdale who is said to have played a leading role in the establishment of the English title registration in the 19th century, "listed six features which should

[93] In his work, Simpson (1976: 17) termed these three sections property, proprietorship and incumbrances. The classification is therefore similar to Nichols' (1993: 7) categorization in splitting information on the people (proprietorship), the nature of the right (incumbrances) as well as on the land itself (property); an overview is provided below.

be combined in a system of registration of title" (ibid.):[94] (1) *security*; (2) *simplicity*; (3) *accuracy*; (4) *expedition*; (5) *cheapness*; (6) *suitability to its circumstances*; and (7) *completeness of the record.*

1. *Security* relates to all transactions and the holding of sticks of the bundle of rights, i.e. leases, purchases, easements and other use-rights and is thus "the quintessence of the system" (ibid.);
2. *Simplicity* refers to the ease of the people to accept the process of land registration, and hence targets on the language used, the terms introduced and forms presented easy to understand;
3. *Accuracy* as well (4) *expedition* are characteristics usually apt to any registry, "for plainly an inaccurate register would be worse than useless" (ibid.), whereas both play an important role as "obvious operational necessities in any system if it is to be effective" (ibid.);

5. *Cheapness* is a characteristic which can only comparatively be assessed in relation to possible alternatives (ibid.: 17–18);
6. S*uitability to its circumstances* is dependent on its feasibility that "obviously depend on the availability of money, manpower and expertise";
7. *Completeness of the record* is achieved either sporadically or systematically, whereas both need to be discussed. However, the registry should compile a complete record of all possible land, for "…until it is complete, unregistered parcels will continue to be intermixed with registered parcels, with different laws applying to each, and therefore important benefits which should accrue from registration of title will not be obtained" (ibid.: 18).

A situation of a *frozen title* exists where the so-called input documents to serve as a proof of title "were not accepted and the old situation was maintained in the register…" (Bogaerts & Zevenbergen 2001: 330). An exception to the completeness of registries applies to so-called *overriding interests*, i.e. "certain rights and liabilities affecting land which it is not practicable to register but which, though not registered, must nevertheless retain their validity" (ibid.). This e.g. may apply to building regulations or particular health care in one area and hence should not be registered in each single parcel situated within this area (ibid.). The author refers to two different sorts of overriding interest: The first are "[r]*ights which may be ascertained by inspection of the property or by enquiries of the occupier*" (ibid.; italics in the original); these rights e.g. over a short-term tenancy may be revealed on inspection which is not recorded by title, but is proven valid due to a particular land policy which targets on short-term leases and prohibits a premature termination. The second sort concerns "*liabilities arising under statute*" (ibid.;

[94] The following six features are taken from Simpson (1976: 17), whereas Simpson lend the seventh quality from Dowson & Sheppard (1952: 72).

italics in the original), as e.g. land taxes. In addition, "…there is always the possibility of compulsory acquisition for some public purpose. It would scarcely be sensible to suppose that such liabilities could be avoided by failing to register them" (ibid.: 18–19). However, seeing that completeness is targeted for a register, these overriding interests "should be kept to a minimum" (ibid.: 19). However, "[n]o aspect of registration of title [which] has caused more controversy than the relationship of boundaries on the ground to the maps, plans, diagrams (…) and verbal descriptions which are used to define the units of property recorded in the register" (ibid.: 125).[95] It is therefore important that reference marks must be of a reliable, solid and long-enduring character (ibid.: 128).

While the description of landed property was witnessed in former times, it had to be described later on in written documents, delineated e.g. in relationship to adjacent parties, or "effectively enclosed by boundaries of reasonable permanence" (ibid. 1976: 129). When the system of title registration started in England and Australia, two different sorts of *parcel demarcation* had been used:[96] While Australia's Torrens system "relies on pegs or beacons indicating turning points on the boundary" (ibid.: 131) together with a so-called *title plan* which is specifically drawn to depict the titled land, the English system relies on physical boundaries that are shown on topographical maps.[97] The difference is certrainly based on the topography of the land in Australia, which is comprised by huge tracts of land and impossible to get fenced. However, the process of ascertaining rights in land in England – with an explicit requirement to define "precise" boundaries – was difficult due to a lack of evidence: People had lived there for generations, thus "landowners literally did not know the precise position of their boundaries. An immensely difficult, and quite needless, problem in 'sporadic adjudication' (…) was thus created. (…)" (ibid. 1976: 134).[98] As a result, "even if the precise line of the boundary could be determined, the cost of making an exact survey was sufficient to deter most landowners from applying for registration" (ibid. 1976: 134). Due to a reform which introduced "general boundaries", "the exact line of the boundary has been left undetermined" (ibid.: 135), e.g. whether the boundary includes a hedge or a wall, or is just drawn next to it. This system "makes it quite

95 A discussion on the term "boundary" and its different means of marking is provided in Simpson (1976: 125–128).

96 The term *parcel demarcation* refers to both in this context, namely its physical demarcation as well as its verbal description (Simpson 1976: 131).

97 A *topographical map* shows what exists physically on the ground, without showing property boundaries except those that are visible by physical features; a *cadastral map* is divided into units of ownership and does not show any physical characteristics, except those physical objects that represent property boundaries (ibid. 1976: 131). An *ordnance map* where "ordnance" means "artillery" was drawn for military purposes, and where large-scale versions of it exists, "have come to serve all the purposes of a cadastral map" (ibid. 1976: 135).

98 The difference between *sporadic* and *systematic adjudication* is treated further below.

clear where the parcel is situated in relation to certain clearly visible physical features, though it does not require the precise relationship between those physical features and the exact boundary lines to be defined" (ibid.).[99]

With view to the fact that in any social setting some sort of land holding can already be encountered, there is "the difficult task of ascertaining title before it can be registered. Compiling the register is the real difficulty; maintaining it once it has been compiled is simple by comparison" (ibid.: 188; see figure 3-3 below). Thus, for the *process of compiling* a register it is necessary to ascertain forms of existing rights together with their respective form of evidence; *maintenance of existing records*, e.g. the quality of former deed registrations, as well as the acceptance of individualization among the populace and the recognition by customary law play an equally important role (especially where no written documentation of ownership exists).

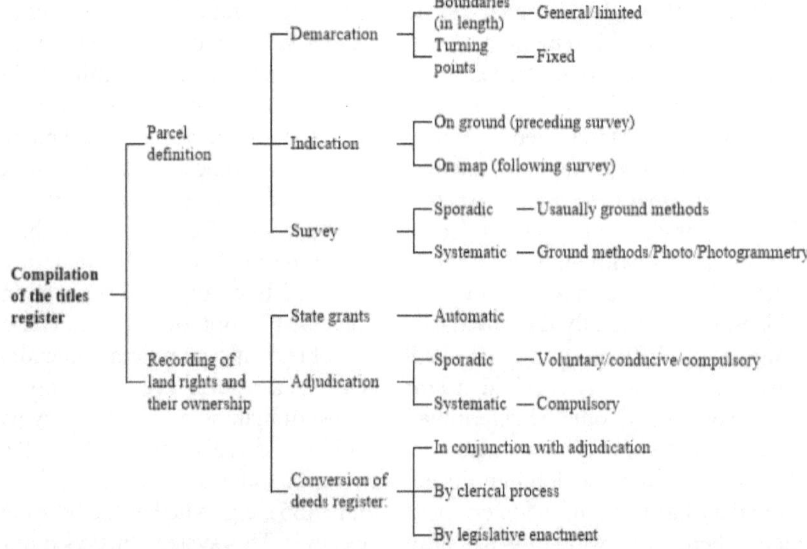

Figure 3-3: Process of compilation (adapted from Simpson 1976: 219)

Before starting to compile any records, a re-planning of the actual pattern of land holdings might be considered, e.g. by *consolidating* dispersed land plots. However, to realize any kind of consolidation – here defined as the restructuring of land holdings in a given area into so-called *units of economic size and rational shape* (ibid.: 246) – "is possible only where compilation is being conducted in

99 The general boundaries rule was a reintroduction of a rule that governed the English system of conveyancing for centuries; it defined six physical features to describe boundaries, namely "[t]he wall, fence, hedge, ditch, stream, and road, which all, except the last, serve the practical function of keeping introducers out and stock in…" (ibid.).

respect of a number of properties at the same time" (ibid.: 189). As is discussed by Simpson (1976: 240), the common definition of the term consolidation usually refers to *fragmentation* of agricultural holdings, which usually conveys a detrimental effect on agricultural production, "for nobody would ever associate 'fragmentation' (or the verb 'fragmentize') with any deliberate or beneficial division of land, or anything else" (ibid. 1976: 240). Thus, consolidation shall refer to the *dispersion* of land plots defined as belonging to the same unit of ownership. It is (Dahlman 1980) who convincingly shows according to the rise and saving of transaction cost on the open-fields in medieval England that dispersed lands perform a particular economic function and are the result of adaptive behavior: Scattering serves a form of risk aversion seeing that "the peasant could be reasonable certain that a random disturbance to the yield on one part of his holdings would not equally effect the other parts scattered in different locations" (ibid. 1980: 59); there was, relative to its cost, no alternative means of insurance. On the other hand, dispersed and, relatively to its size, unproductive plots prevent outsiders and hence lowers policing cost on the fields; it is thus an effective means "to combine private property with collective control" (ibid.: 124).

Compiling the record is made either in a *sporadic* or *systematic* approach; however, means of compulsion must apply to the maintenance of a register where existing titles and their subsequent transactions are being recorded (Simpson (1976: 191).[100] The *systematic compilation* is comprised by defining the unit of record, "the determination of rights and interested parties, and their registration; in a methodical manner and in orderly sequence, district by district, village by village, block by block, parcel by parcel, throughout the territory concerned" (ibid.). This systematic approach is contrasted by a *sporadic compilation* that is defined as "...any process of defining parcels, of determining rights and interested parties, and of registering these effects, which is applied in a piecemeal manner, now here, now there, to scattered parcels over an indefinite and unpredictable period" (ibid.). Thus, due to its "haphazard nature" (ibid.) this sporadic approach is viewed rather critically. As a compromise Simpson (1976: 190) suggests to systematically register in selected areas and (in contrast to a sporadic approach) according to "a process which is effected compulsory throughout an officially designated area of land, (...) whatever the size of the area" (ibid.) Accordingly, the systematic approach is inevitably based on compulsion, whereas sporadic efforts "may be compulsory or conducive or voluntary" (ibid.: 190–191):[101] When *compulsory*, a land owner is compelled to register a title for his holding; when being

[100] This definition is lent from Dowson & Sheppard (1952: 93).

[101] In his classical work Simpson (1976) reveals firstly that only so-called *selective compulsion*, meaning a request to registration of title to be made compulsory by specific order of a county council in specifically defined areas in 19th century England (Simpson 1976: 44), led to the likelihood that "the Central Government was given the power to initiate extensions of the compulsory system instead of the county councils" (ibid. 1976: 46), and he summarizes, "that the English system failed until compulsion was introduced in 1925"

conducive, "registration is induced merely because serious disadvantages result from failure to register" (ibid.: 191).[102] In contrast, a *voluntary* system is offered where the choice to register is left optional (ibid.). Trebilcock & Veel (2008: 481) see the strongest argument for a voluntary and sporadic approach "that it avoids the myriad unforeseeable and potentially negative consequences that can result from the top-down imposition of a uniform system of property arrangement". On the contrary, seeing that "people are generally anxious to have their rights registered and only nominal fees are charged, normally almost all parcels in an area can be adjudicated in a single field operation" (Bruce 2006: 29). In particular, the advantages of a systematic approach "are seen as best able to take advantage of economies of scale in measurement, adjudication, and conflict resolution" (Bledsoe 2006: 165). Thus, Simpson (1976: 191) highlights that "...without compulsion in its introduction and effective inducement in its maintenance, registration of title will not succeed. The question is not whether there should be compulsion but to what extent compulsion is feasible" (Simpson 1976: 191).

The *components of compilation* consist of two parts, namely to collect information on (A) the parcel, and (B) its owner: "The former results in a map or plan; the latter in the register itself" (ibid.: 192).

A) For the identification of the parcel, three distinct steps are necessary, namely (i) *demarcation*, (ii) *indication* as well as the (iii) *survey*: (i) *Demarcation* is the process of physically marking the ground with boundaries to be recorded; if the terrain is not demarcated on the ground, so-called *imaginary lines*, or *monuments*, i.e. "any tangible landmark indicating boundaries" (Dale & McLaughlin 1999: 50), are used which may be either specific, so-called *fixed*, or *general boundaries* (ibid. 1999: 50): The first denotes boundaries already determined and defined on the ground, while the latter refers to approximate lines that have not yet been adjudicated on-site. However, "[t]he accuracy of survey measurements in specific boundary systems is of secondary importance; monumentation and defined limits

(ibid.: 89). This "effective measure of compulsion" (ibid.: 71) was directly built-in to the system of Torrens in mid-19th century Australia. However, an institutional change toward the new system did not bring the expected results, as Simpson writes (ibid.: 72): "Anyway, whatever the reason, the advantages of the Torrens system did not suffice to bring in by voluntary methods the old titles which had been granted before the system was introduced. More than a century later the two systems still exist side by side in all the Australian States". Two main reasons are mentioned for these missing effects: On the one hand, it is stated that the Torrens system did not work as a system of insurance as it was the case in England (ibid.: 80). On the other hand, the Torrens system did not succeed in England especially vis-à-vis its first registration, as long as it remained voluntarily (ibid.: 195).

[102] The English process, as presented shortly above, is known for its "compulsory first registration" (ibid. 191). In those "compulsory areas" a sanction is imposed against those purchasers who fail to apply for registration (ibid.).

of possession on the ground are the primary considerations" (ibid.: 51);[103] (ii) *indication* is the official positioning of a parcel's boundaries; it follows (iii) the *survey* of the marked boundaries (so-called *preceding survey*) together with "the preparation of a map or plan illustrating them" (Simpson 1976: 192; so-called *following survey*). For the purpose of identifying the parcel, so-called *cadastral surveying*, geometrical data must provide "...the information necessary for the overall planning and coordination of land titling projects, for assessing evidence in support of the initial registration of title, for the actual delimitation and registration of boundaries" (Dale & McLaughlin 1999: 46). The authors stress that those "...also form the basis for the subsequent re-establishment of boundaries in the case of uncertainty, dispute, or subdivision" (ibid. 1999: 46). Surveying is thus viewed as "an investment in the future to ensure the long-term maintenance of the parcellation of the land" (ibid. 1999: 52). The costs though depend "on the precision that is sought – that is how precisely the boundaries are defined and must be measured" (ibid.: 52). The choice for appropriate measurement standards and methods, as well as the requirements on the staff carrying out the survey work, e.g. licenses, is usually prescribed by law and regulations and depend on the local circumstances; in many cases, the monitoring of these standards is taken over by a central government agency which is then responsibility for the work's precision (ibid.). There are broadly two different categories of *survey techniques*, namely field surveying and photogrammetry, as well as two *categories of output*, graphical and numerical (ibid.: 54):[104] The traditional surveying technique is based on ground methods, whereas notes by the surveyor are either recorded in a field notebook or, more recently, electronically; photogrammetry is based on aerial photography whose measurements are mostly recorded in an office; both techniques may comply with "the necessary standards of accuracy and precision for most land administration purposes" (ibid.). Surveying work based on photogrammetry is cheaper, but constrained though for it "cannot (...) be used for setting out, other than as a basis for the design of plot layouts, and its use is dependent on the existence of property boundaries that are visible from the air" (ibid.). Similar constraints apply to the usage of satellite imagery and remote sensing (ibid.).

Besides, these "issues are those of cost and effectiveness under the circumstances, and the legitimacy of such techniques under regulations that have been designed for surveyors working on the ground" (ibid.). The advantage of aerial methods is moreover the production of a historical record which may later be compared with those produced in the future "to see what changes have taken place and even to re-measure conditions in the past. Thus where disputes arise over

[103] Accordingly, Dale & McLaughlin (1999: 51) stress that the boundaries do not need to be of permanent nature, hence can alternatively be based e.g. on agreements among neighbors or adverse possession.

[104] With view to the legitimacy of the data produced, "relative weight given to different type of evidence varies between jurisdictions. In most countries the marks found in the ground take precedence over the numerical values" (ibid. 1999: 51).

whether a boundary has been moved, old aerial photographs can provide crucial evidence" (ibid.). Survey work on the ground was mostly realized by using theodolites, steel tapes or electronic distance measuring systems; today, "[i]ncreasing use is being made of global positioning systems (GPS) techniques that can provide coordinate values for points on the ground to a high level of precision" (ibid.). These high-precision techniques though depend on time and a relative expensive equipments. With view to the lacking of financial means in so-called less developed countries the choices are constrained "since modern technology must either be donated or else be paid for from very limited amounts of hard currency (ibid.: 56). The authors thus emphasize that "[w]hat is however of primary importance is the end product and not the technology that is used to achieve that end" (ibid.). (Dale & McLaughlin 1999: 46) stress that "[s]imple and inexpensive techniques are often just as effective as more expensive and sophisticated methods".

Traditionally, the initial registration of property is taken over by state officials to conduct the adjudication process and cadastral surveying; in more recent times, "private sector surveyors have carried out some of the work on contract in an attempt to reduce actual costs and to accelerate the pace of projects" (ibid. 1999: 46). Subsequent surveying of changes to the boundaries are either realized by the public sector or private licensed surveyors (ibid. 46).

The costs are either covered by the state or through aid projects (ibid.). With view to the first, the financing of land administration systems may be based on three different sources, i.e. by *taxes*, *fees* or *commission* (ibid.: 139): *Taxation* establishes no direct link between the activity that raises the tax and land administration, "there is thus no direct incentive to optimize the service delivered" (ibid.); in case of financing by *fees* the government decides over tariffs, and to some extent users pay for the delivered service while payments are directly or indirectly channeled to the agency; finally, "[f]inancing by *commission* means that an applicant pays for the service and that the agency has the authority to decide about the tariff based on rules set by the government" (ibid.). The sources of financing depend on policies and specifically on the government's stance toward the organization and (1) its degree of independence to decide about its income, active marketing and expansion; (2) its activities to raise money for investments; (3) its accounting system; (4) its system of remuneration; as well as vis-à vis (5) its internal governance structure (ibid.: 139–140).

B) The (interest of the) owner and the recognition of any possible subordinate interests "...affecting the ownership must be ascertained at one and the same time; claims to ownership and claims to subordinate or derivative interests must be considered together if anything relevant is not to be overlooked" (Simpson 1976: 193). Here, three processes may become relevant (ibid. 1976: 193): (i) *registration* of State grants; (ii) *adjudication* of existing rights; (iii) the *conversion* of deeds. These steps are profoundly discussed as follows.

(i) The *registration* of State grants is considered "[t]he nearest approach to a clean sheet in land holding (...) when the State disposes of land which it considers to be at its disposition by virtue of statute or its own prerogative" (ibid.: 193): "In such cases, without any process for ascertaining whether any rights exist, the State grants a title which is absolute (or indefeasible or unimpeachable — whichever word is preferred to denote this particular quality) " (ibid.). A registration based on such an initial record is thus easy to realize, which then only demands subsequent maintenance of the record (ibid.).

(ii) *Adjudication*, is defined as "the authoritative ascertainment of existing rights" (ibid.) that is either *sporadic* or *systematic*.[105] However, the authors put emphasis on the "cardinal principle of adjudication that it does not, by itself, alter existing rights or create new ones. It merely establishes with certainty and finality what rights exist, by whom they are exercised, and to what limitation, if any, they are subject" (ibid.195). Besides, adjudication is advised if a government intends to grant or use land that is currently not occupied (ibid.).

The benefits of *systematic* adjudication are discussed as follows (ibid.: 202–204):[106] Accordingly, systematic adjudication...

- ensures public awareness and is therefore "...a very effective safeguard against fraud or concealment or even honest oversight. Systematic adjudication alerts the whole neighbourhood..." (ibid.: 202); however, seen on the English example it could be learned that "this very publicity would be considered a disadvantage by those brought up in the English practice which, almost alone in the world, still favours secrecy in land dealing" (ibid.); moreover, there can be trouble provoked in the intimate neighborhood as well as beyond, if consent on borders cannot be achieved (ibid. 204): "Indeed, in countries with a well-established system of documentary title and effective private conveyancing the upheaval which systematic compilation would cause is generally considered a sufficient reason for not attempting it";
- is "carried out on or near the actual ground" (ibid.), which "can hardly be exaggerated" (ibid.), e.g. to inhibit officially transferred misinformation with

[105] The term *systematic adjudication* of land was initially known as *settlement* – in the sense of fixing a specific amount of land for fiscal purposes in a defined area – a process that started 1789 in India while "title was only incidental and sprang from the presumption that he who paid the tax was the owner" (Simpson 1976: 194). Due to juridical confusion with the meaning of "land settlement", i.e. settling families on land, or vice versa, settling land on families, the term was changed to systematic ascertainment (ibid. 1976: 194).

[106] The following steps are necessary for carrying out a systematic adjudication process ((Dale & McLaughlin 1999: 49): (1) Issue a law and make it public; (2) select areas with a given priority vis-à-vis relevant dealings in land, need in using land as collateral, frequent occurrences of litigation, land use changes or changes in land holdings, as well as where "lack of certainty inhibits development and, in practice, it is politically expedient" (ibid. 1999: 49); (3) give maximum publicity to the areas concerned; (4) appoint appropriate staff; (5) realize the adjudication process on the ground; (6) calculate a period of time to proceed with "objections, appeals, and the rectification of the initial adjudication where appropriate" (ibid.: 49), and compile the information in the title register.

view to "all the local inhabitants who have knowledge of the past history of the land and may be able to supply important information at any stage of locally conducted investigation" (ibid.: 203); on the other hand, this raises cost, as is expressed by an English registrar in 1971 who is convinced about the suitability of a sporadic approach, seeing that

> ...the title to the unregistered land has been deduced by the vendor's solicitor, examined by the purchaser's solicitor, and usually – almost certainly in the case of dwelling houses – examined by the mortgagee's solicitor, so when the title comes to me I can save a great deal of expense by examining that title cursorily instead of meticulously... (ibid.: 204)

- is administratively facilitated, seeing that personnel and general overhead expenses are foreseeable, appropriate staff is on site and the respective "flow of work can be more competently regulated" (ibid.: 203);
- eases survey work: whatever techniques used to survey the ground, "[c]ommon boundaries are surveyed in systematic survey once only and this is much cheaper than ad hoc survey for individual landowner" (ibid.);
- facilitates the re-planning of land holdings, if e.g. the consolidation of dispersed land plots is intended;
- encourages the introduction of "a single uniform law" (ibid.) that applies to all land at the same time, to avoid a situation where "two or more different laws be applied side by side as they are when registered and unregistered parcels are indiscriminately intermixed in the same locality" (ibid.);
- and, finally, "enables registration to be completed in those areas which matter most. Without systematic adjudication there is no real prospect of completing the register, and the benefit to the owner of his registered title will be largely lost (ibid.: 203–204).

On the contrary, the benefits of *sporadic* adjudication are discussed below (ibid.: 204–207): Sporadic adjudication...

- is – as its greatest advantage – voluntarily and only proceeded if a party is interested (ibid.: 204); it is thus "market-driven but does not allow for the comprehensive collection of land records that may be beneficial to the community as a whole" (Dale & McLaughlin 1999: 49). Moreover, this procedure takes more time, and carries with it "the danger of encouraging the registration of only weak titles where conveyancers are unwilling to provide a warranty for transaction; good titles then tend to become those that have not been registered" (ibid. 1999: 49).[107]

[107] This phenomenon of uncertainty based on asymmetric information is also known as a "market for lemons", see Akerlof (1970).

- saves costs to the public, for it "enables the financial burden to be transferred to the individual, whilst systematic adjudication must initially be paid for by the State" (ibid.: 205). However, it gives advantage to "an owner who has no intention of dealing and no fear that his title is insecure, and who therefore appears to gain nothing from registration" (ibid.);
- does not afford an appropriate number of staff, since the process – if there is no apparent reason against it – may only take longer; by contrast, Simpson (1976: 205) rails against this argument seeing that "it is systematic adjudication that can be precisely adjusted to the situation; sporadic adjudication, even if it is possible to forecast its incidence, cannot so easily be turned on and off";
- supports "those progressive farmers who had broken the barrier between subsistence cultivation and commercial farming" (ibid.); this argument though assumes that "in the long term more might be gained by winning over the support of these [social and political] leaders" (ibid. 206–207); however, as evidence shows (e.g. Simpson 1976: 206–207; Sikor & Müller 2009; Sjaastad & Cousins 2009), such an approach supports rent seeking by those farmers having been beneficially "selected" to registering rights to land "before other landholders realize the significance of such claims…" (Simpson 1976: 206); besides, the expected outcome did not materialize with view to other members still exercising customary rights, hence resulted in the fact that these got isolated from those applying customary law systems. Consequently, it seems reasonable that a systematic approach "with full group understanding and participation is more likely to overcome these difficulties…" (ibid.); a selective principle applied to only a few farmers is, moreover, "inconsistent with the declared principle of recognizing the existing interest" (ibid.).[108] The procedure of adjudication must however "be applied equally to all and it would be unjust to allow recognition of one owner, a good farmer, and refuse another because, for instance, his holding was too small. It might well be the only land he owned (ibid.).

Finally, (iii) the possibilities for the *conversion* of a deeds register differ greatly, seeing that every country

> …which decides to introduce registration of title already operates some system of registration of deeds, and not only is there a very wide variation in systems of deeds registration, but the use made even of similar systems has also varied greatly where conversion is concerned. (ibid.: 207)

Simpson though stress that "…there is no bigger obstacle to the introduction of registration of title than the existence of an effective register which appears to be satisfactory and which, naturally, there is reluctance to change" (ibid.: 212). A

[108] This quote is lent from the Kenya Party Report on African Land Tenure (1958: 19).

few examples shall illuminate these circumstances:[109] In Australia, where the Torrens system has been introduced on a voluntary basis a duality together with customary (Aboriginal) land tenure still exist today (Brazenor et al. 1999), the first possibility is that a state grant is "automatically registered on issue" (Simpson 1976: 208). By comparison, in Singapore, where Torrens' provisions are equally exercised, it is a copy of the grant which is issued and which serves as certificate of title. Secondly, another option is to apply for registration on a voluntary basis which is rarely being used due to its high expenses and the certainty provided by the traditional deeds system. The third way is a so-called selection by the Registrar among those titles that are subject of a dealing. The compulsory system introduced in England requires that "all titles have to be registered whenever there is a sale" (ibid. 1976: 208). In other countries, e.g. South Africa, deeds registrations are regarded equal to registration of title (Sjaastad & Cousins 2009: 4). In general, however, where a deeds register is transformed into a title register the process is based on identifying the necessary location and veracity of the information to be registered (ibid.). …the procedures for adjudication begin by identifying where the data is located and then checking their veracity (also on the ground) before registering the title, e.g. due to inheritance (Dale & McLaughlin 1999: 48).

Thus, Simpson (1976) establishes four categories "in respect of deed registrars and their possible conversion to register of title" (Simpson 1976. 212): With regard to the example of South Africa, a first group of countries does not intent to make any changes to the deeds system but equate it with title registrations, or, as do India and Pakistan, regard their fiscal land records equally. A second group, presented e.g. by Sudan and Zambia, just changed their land law to convert deeds into title registrations. A third group, as e.g. Germany or Malaysia, used their deed register (or the cadastral registers in the case of Germany) and converted it into a provisional register where "the onus is not on the Registry to produce the perfect title (though it should do its best) but on the proprietor to challenge what he does not accept" (ibid. 1976: 212). The fourth group, as e.g. Kenya or Turkey, use the provisional nature of their deeds register "to assist in or supplement the process of adjudication…" (ibid.: 212). This latter procedure is given, specifically with regard to title on first registration, support by highlighting that compiling a register "can be greatly facilitated, as well as greatly accelerated by calling on the aid of time to test and validate initially uncertain results" (ibid.: 212–213). Accordingly, the English Land Transfer Act of 1875 introduced three sorts of provisional title which were legally binding: *Absolute* is the title if it is "final, conclusive and guaranteed" (ibid. 1976: 215); a so-called *qualified* title is of the same nature "except for some particular blemish which is specified in the register" (ibid.: 215); *possessory* title "is given to an applicant who has actual possession of the land in question, or receives its rents and profits, but has incomplete documents of title (ibid.). However, Simpson highlights that "…the particular virtue of this sort of

[109] If not mention otherwise, the next examples are taken from Simpson (1976: 208–213).

title is that in due course it ripens into an absolute title, because, after the lapse of fifteen years in the case of freehold (or ten years in the case of leasehold), the Registrar is bound to convert a possessory title into an absolute title, provided that the applicant has proved that he is still in possession of the land" (ibid.). A fourth title is provided in respect to leasehold, so-called *good leasehold* title, defined as "the same as an absolute leasehold title except that the right of the lessor to grant the lease is not guaranteed which, of course, means that the freehold title is unregistered" (ibid.). Accordingly, the Common Law provides for *adverse possession*, i.e. "the peaceful and unchallenged occupation of land" (Dale & McLaughlin 1999: 48) that may be transformed to absolute ownership after a specific period of time. Hence, the registration of rights may also start with existing unregistered deeds and other documents, so-called *first registration* in England and Wales (ibid. 1999: 48).

Thus, registering land is, in short, "the process of making and keeping records of property rights" (Dale & McLaughlin 1999: 10), the analysis focuses on the questions of how property rights did emerge and vanish at various points in time, and how the bundle is held over time (Werin 2003: 14–15). Besides, it is important that the process of registering land is being…

> …kept in perspective. It is a device which may be essential to sound land administration but it is merely a part of the machinery of government. It is not some sort of magical specific which will automatically produce good land use and development; nor is it a system of land holding; (…) it is not even a kind of land reform, though it may be a valuable administrative aid to land reform. In short, land registration is only a means to an end. It is not an end in itself. Much time, money, and effort can be wasted if that elementary truth be forgotten. (Simpson 1976: 3)

3.3 Institutional change of land tenure

The relationships between people and land which is regulated by the respective, so-called agrarian constitution (*Agrarverfassung)* which depend on natural circumstances, e.g. soil quality and climate, social traits, as the legal framework, ideology or the social and economic structuring, as well as exogenous events, as (neo-)colonialism (Kuhnen 1982: 69). The term Agrarverfassung may be translated as *land tenure* that comprises the total of relationships between people and the (bundle of) rights to control and use land (ibid. 1982: 69). As such, the Agrarverfassung is part of the overall legal and social structuring in society which contains the property constitution (*Grundbesitzverfassung)* and labor constitution (*Arbeitsverfassung*) and is hence of essential importance for society (ibid.: 69).

The first defines the exact relationship between people and land. With view to the fact that land cannot be multiplied but is the most cost-saving source of agrarian production (see chapter 5.2.2.2), the quantity and quality of the soil shape social conditions of agrarian societies, while ownership structures determine "the distribution of labor and income, individual status, social strata, as well as power and influence" (ibid.: 70; see chapter 3.1.4). The concept of private ownership e.g. is historically a Western concept which has been introduced by Europeans in the so-called developing countries; other property arrangements are based on alternative institutional structures that regulate land use (ibid.: 71). While, from a Western point of view, private ownership is said to provide security for investments, an increase in population (or other reasons that make land to become a scarce resource) without taking measures for alternative settlement structures "leads to losses of ownership of the poor and simultaneously a concentration of wealth among the rich, and thus diminishes any positive effects" (ibid.). By keeping most of the land in state-ownership and organizing agrarian production plan-based the Soviet Union in contrast prevented the exploitation and unearned income from ground rent which originate from ownership of non-reproducable land resources (ibid.: 72). In general, the reasons for keeping land in state-ownership, whether governed on local or governmental level, might either depend on land use, relate to ideology, or if social concerns would not be met by private property, e.g. in the case of state forests or border zones (ibid.). The Grundbesitzverfassung's form of appearance is hence "the result of historical, political, cultural and economic development" (ibid.: 72), and might therefore be modified by a change of political ideas, population growth (or decline) as well as technical or economic changes (ibid.: 70). On the other hand, alterations might be blocked by political and economic (power) structures, legal rules which e.g. relate to inheritcance, contractual arrangements in production, as well as religious or ethical concerns (ibid.: 70).

The Arbeitsverfassung, on the other hand, was formerly based on family farms (so-called *Familienarbeitsverfassung*) and have been for centuries the most common form of labor organization in agriculture (ibid.: 76): The family produces for their own needs and is in general exclusive, which increases her direct interest for (and sets incentives to) maximizing production. This organizational mode is challenged though by a fluctuating and notably declining number of the (available) working force within the family, which is balanced by leasing land in or out, by employing further personnel to sustain the unit of operation, or by the availability of alternative off-farm employment; however, with increasing technical development in agriculture the family farm can hardly compete with the specialization of labor (ibid.). The number of landless agrarian workers has thus been increasing worldwide where no alternative forms of employment exist (ibid.). Other forms of agrarian organization (summarized under a so-called foreign employment constitution (*Fremdarbeitsverfassung*)) are comprised by temporary and permanent labor arrangements (ibid.: 76–78).

Seeing that both Grundbesitz- and Arbeitsverfassung determine land use, land distribution and mode of production, they form a peculiar argrarian system which is marked by specific institutional, social and economic structures that follow particular economic and social ethics, e.g. pastural systems as transhumance, feudalistic systems, family farms as well as collective or capitalist agriculture (ibid.: 78-

83). The transformation process from a plan-based economy to a capitalist system thus challenges the agricultural sector (and as such the formation of a newly created agrarian constitution) vis-à-vis its contribution to the overall ecnomy and, in particular, the further development of the formerly collectively organized farms, ecologically friendly land use and livestock husbandry, as well as an appropriate regulatory framework that gives the relevant politicians and organizations the means to resolve potential conflicts about the future path of the agricultural policy (Hagedorn 1992: 191). The political course is hence, as was mentioned above, not only led by socially acceptable ideas but rent-seeking activities of the relevant interest groups to benefit from the expected restructuring of ownership titles and claims (ibid. 1992: 192). The agrarian constitution hence depends, on the one hand, on (opportunistic behaviors to be found in) political processes, e.g. electoral control and party competition, interpretation systems to legitimizing agricultural policies, collective action of interest groups and bureaucracy; and, on the other hand, (the evolution of) transaction cost minimizing institutional arrangements, such as product and factor markets, family farms as well as institutions to internalize external effects, e.g. for environmental protection (ibid.: 193).[110] As is shown thoroughly by Hagedorn (1996: 288–300), it is the family-based agrarian constitution who most probably succeeds in countries that formerly belonged to the Soviet Unon, as the family manages to keep transaction costs low, while its integrative character offers a high degree of flexibility, and enhances motivation and the willingness to cooperate among its members (ibid. 1992: 194–195).[111],[112] As a consequence of technical and institutional changes though farming became more centralized and is characterized by vertically integrated organization of large farms which operate among less market participants but produce most of the output (Hagedorn 1996: 54). Besides of the pressure to adjust to structural changes in agricultural production further conflicts stem, on the one hand, from new information technology, innovations as bio-food and fuels or genetically modified organizms as well as, on the other hand, external effects of agrarian production or, e.g. vis-à-vis European integration (ibid. 1996: 44–55). However, Hagedorn (1996: 44) emphasizes that the innovation sequence is foremost motivated by po-

[110] Another leading factor that influence the type of agrarian constitution stems from the interest-led role of agriculture in the overall political process, see Hagedorn & Schmitt (1985).

[111] A rising number of employees might, up to a certain point, reduce production cost, but lead to an increase of transaction costs due to the rise of administration and monitoring cost to counter e.g. free-riding behaviors (Hagedorn 1992: 194–195).

[112] By operating as a market-independent economic entity, the family unites the three production factors, i.e. land, labor and capital, as sources of income which can be easily reallocated, while its integrative decision and intra-family coordination structure (the family as a social system, the interdependent economic operation of the household and the farm, and the family farm as technical productive entity) holds transaction cost low which make the family farm an adaptable and thusly robust economic organization (ibid. 1996: 335).

litical interest. E.g. the so-called "urban bias" explains the economic discrimination of the agricultural sector by the policital elite in rather traditionally organzied (agrarian) societies (Lipton 1977).

3.3.1 Features of institutional change

To better understand the nature of institutional change and its consequences Streeck & Thelen (2005) distinguish between the *process* of change and the *results* that the change brings about (Streeck & Thelen: 8–9); see table 3-4 below): The *process* is classified according to as either abrupt and sudden, or incremental and gradual; the *results* can lead either to continuity or discontinuity. It is assumed that the response to an incremental change is "fundamentally reactive and adaptive and serving to protect institutional continuity" (ibid. 2005b: 8; upper left cell of table 3-4); on the contrary, discontinuity follows from abrupt institutional breakdown and by replacement (lower right cell).

		Results of change	
		continuity	discontinuity
Process of change	incremental	reproduction by adaptation	gradual transformation
	abrupt	survival and return	breakdown and replacement

Table 3-4: Types of institutional change: processes and results (Streeck & Thelen: 9)

In real life though the authors stress that on the one hand "there often is considerable continuity through and in spite of historical break points" (ibid.: 8–9) which actually result in institutions' survival and return (lower cell on the left); on the other hand, a gradual transformation (upper right cell) might in effect be expected in case of "dramatic institutional reconfiguration beneath the surface of apparent stability or adaptive self-production, as a result of an accumulation over longer periods of time of subtle incremental changes" (ibid.). *Abrupt* or sudden changes, stemming from exogenous shocks or crises as wars, revolutions or earthquakes, may disrupt the status quo and promote fundamental institutional changes, so-called *critical junctures* (Campbell 2010: 92).[113] On the other, so-called *tipping points* lead to far-reaching change "through the accumulation of small, often seemingly insignificant adjustments" (Streeck & Thelen: 8).

3.3.1.1 Incremental change and path dependency

With reference to North (1990: 6), the typically *incremental* or evolutionary nature of institutional change stems from the embeddedness of a society's informal

[113] Critiques on the shortages of the concept of critical junctures is found e.g. in Streeck & Thelen or Mahoney & Thelen (2010).

constraints (Granovetter 1985), an idea that is rooted in *path dependency*, a concept closely related to institutions. It views the latter as "sticky, resistant to change, and generally only change in 'path dependent' ways" (Campbell 2010: 90):[114] It is a process characterized by long-lasting institutions which are the result of decisions or contingent events and which may hence constrain choices available in the future (ibid. 2010: 90). Underlying the concept of incremental or gradual institutional change are two complementary processes (Lund 2006: 687): On the one hand, the process of regularization which is understood as "processes which produce rules and organizations and customs and symbols and rituals and categories and seek to make them durable" (Moore 1978: 50); on the other hand, the process of situational adjustment describes opportunities where people generate and/or exploit a situation's indeterminacy either by reinterpreting or by redefining rules and relationships. While the first "is the result of people's efforts to fix social reality, to harden it, to give it form and predictability" (Lund 2006: 699), the latter is about the "manipulation of rules and manoeuvring between them impute a measure of unpredictability, inconsistency, paradox and ambiguity, and ultimately institutional incongruence" (ibid. 2006: 699.). Usually both kinds of processes are at work at the same time "but only by detailed examination of the outcomes of institutions' acts of governance can a broader aggregated picture be established" (ibid.: 699). A change in rules in the political sphere thus may take place at a slower pace due to the following reasons:[115]

- Slower pace in change might be related to the cost to set up an institutional arrangement; once it has been established, any change is hardly feasible.
- Some institutions are designed in a way that makes it difficult to change them, e.g. due to specific procedural rules.
- With view to the fact that people got familiar with an institutional arrangements people are not keen to change them again.
- Those who benefit of an institution will reinforce their behavior as long as they are continuously provided with these benefits.
- Seeing that the interdependence of a set of institutions result in synergies, any changes in one institution might alter another that "undermine the benefits resulting from institutional complementarity" (Campbell 2010: 90).

Campbell thus emphasizes, "the institutional configuration of different types of national political economies tend to be rather stable even in the presence of considerable pressures for change" (ibid.; Aoki 2001; Crouch 2005: 30-1).

Based on this approach Sehring (2009) shows how institutions in former Soviet countries reacted differently to their new circumstances. He warns though about

[114] The concept of path dependency is further treated extensively in Ebbinghaus (2005), Djelic & Quack (2007) or Mahoney (2000).

[115] The following enumeration is taken from Campbell (2010: 90).

the difficulty inherent in the concept of *path dependency* to get trapped into the ideas of historical determinism where historical processes are rather seen as static and "path dependence arguments are often very deterministic" (Campbell 2010: 92; Ebbinghaus 2005; Haydu 1998). In contrast, past experiences are seen as a continuous process where "behviour or identities that once proved to be successful and that are established, will be used again to meet new challenges" (Sehring 2009: 64).[116] Though paths are not determined by past experiences, further studies in Eastern Europe after the demise of the Soviet Union show that "[p]ath-dependency suggests that the institutional legacies of the past limit the range of current possibilities and/or options in institutional innovation" (Nielsen et al. 1995: 6). In other words, "the concept is to stress the *limited degrees of freedom* that exist for innovation, even in moments of extreme upheaval" (Streeck & Thelen: 6).[117]

3.3.1.2 Induced institutional change and path dependency

In cases where new institutional patterns are adopted by analogy with existing norm Douglas (1986, 1973) argues for that rules must get "naturalized". This view gets support by evidence of comparative analyses, e.g. in case of attempts to liberalize (and change) rural market institutions. It is assumed that liberalization and market penetration as a result of higher competition "will reduce the extent to which such labour relations are based on custom and non-market obligations such as, for example, patron-client relations" (Peters et al. 2012: 17). In contrast, evidence shows that "social institutions are being refashioned by market exchange, becoming more economic in their content and roles, but still shaping economic action in ways which are quite distinctive to these institutions" (Harriss-White & Janakarājan 2004: 158). Consequently, Peters et al. (2012) detect "a muddle of institutional multiplicities" (Peters et al. 2012: 17) as a result of interventions by the international donor community which "simply add a new layer of rules, without overriding others" (Di John 2008: 33), where "individuals and organisations appear to operate often simultaneously in multiple institutional systems, governed by very different sets of incentives" (ibid. 2008: 33). Thus, with regard to the universal, rather normative goals of the international donor community that aims recipients to catch up with "development" and become welfare states as their "counterparts" in the West, it cannot be expected that these prescriptions lead to homogenous paths: "That would be to deny path dependency and to be insensitive to the different ways in which societies and geographic zones are presented within globalization, and as a result are able to construct different welfare mixes" (Wood

[116] This assumption stems from the expanded concept of 'institutional bricolage' which acknowledges the interplay between gradual institutional change and path dependency (Peters et al. 2012: 16).

[117] The authors highlight that "the characterization of change as 'path-dependent' is understood as a refutation of and an alternative to voluntarist ('rational design') accounts that view institutional-building as a matter of constructing efficient incentive structures on a more or less 'clean slate'" (Streeck & Thelen: 6; Stark .

& Gough 2006: 1710). The topic is subject to the endogenous institutional change model by North (1990) and others (see chapter 3.2.3 below). Why and how institutions change is subject to the following part.

3.3.2 Theories on institutional change

Broadly termed there are three "classic" approaches to institutional change, i.e. the traditional property-rights school, the induced institutional change model and North's model on endogenous institutional change. The following part sheds light on their conceptual underpinnings as well as their limits; the final part thus merges these thoughts into a final definition to capture and understand the institutional changes of property rights in agricultural land, and their effects on Georgian agricultural production.

3.3.2.1 Property-rights school

The so-called property-rights school is associated with works of (Coase 1960), Demsetz (1967) and Posner (1977) who understand a change in individual behavior as response to a change in cost-benefit structures (Bromley 1989b: 12), for "...property rights develop to internalize externalities when the gains of internalization become larger than the cost of internalization" (Demsetz 1967: 350). It is the exclusive assignment of private property rights which "provide a direct incentive to improve efficiency and productivity, or, in more fundamental terms, to acquire more knowledge and new techniques" (North & Thomas 1977: 241).[118,119] This stream of thought hence "is primarily concerned with the level of transaction costs that arise from jointly held assets" (Bromley 1989: 14). Seeing that numerous studies have shown that the individualization of property rights is not the *condicio sine qua non* of resource use (see e.g. (Dahlman 1979; Ostrom 2008; Bromley 1992; Agrawal 2001), the approach is confronted by the fact that the most economically appropriate structure of property rights cannot serve as an explanatory variable; in contrast, the nature of ownership rights, whether it is state, private or communal property, is a dependent variable defined by the "function of the economic surplus available to support those differential cost" (Bromley 1989b: 16), and thus relates to (i) the technology available for production, (ii) the state of the market, i.e. the nature of output, demand and relative prices, (iii) resource endowments and relative factor prices that shape the technology of production, as well as (iv) transaction costs, "the usually forgotten element that turns out to be so crucial, for there is no unique ranking of transaction costs with respect to the

---

[118] This notion is taken from North & Thomas (1977) who relate the "first economic revolution" to the superiority of private ownership rights. North, however, later relaxes his quantitative concept on the evolution of private property rights (his works from 1981) toward an institutional approach, see North (1981; 1990).

[119] One of the authors refers to conditions as *efficient* "where the existing set of constraints will produce economic growth" (North 1990: 92).

three classes of ownership" (Dahlman 1980: 204).[120] Underlying this stream of thought is that if resources become scarce and excludability is granted, some form of ownership will emerge; collective ownership though, if well defined, "can be quite compatible with efficient resource allocation and private wealth maximization" (ibid. 1980: 204). On the contrary, the specific nature of the production process and its related exchange costs are "the key to the choice of property (and other) institutions" (Bromley 1989b: 17). In other words, "[t]he nature of transactions costs must be specified in detail before we can account for the choice of property rights in any particular context" (Dahlman 1980: 139).

3.3.2.2 Induced institutional change model

The second model of induced institutional change is primarily rooted in Ruttan and Hayami's (1984) article "Toward a theory of induced institutional innovation". The concept is based on the idea that changes in resource endowments and technical change "induce changes in private property rights and in the development of non-market institutions" (Ruttan & Hayami 1984: 203; Anderson & Hill 1975). It shares the basic features of institutions as defined in this study, such as forming expectations and providing order in society (ibid. 1984: 204), but views institutional change through a neoclassical lens (ibid.: 205): A rising demand for institutional innovation (driven by changes in relative resource endowments and product demand which leads, similar to Marx' view on institutional change, to technical change) is confronted with the supply of institutional change "through the impact of advances in social sciences knowledge and of cultural endowments" (ibid.). In contrast to Marx' vision of being necessarily a dramatic or revolutionary process,

> ...basic institutions such as property rights and markets are more typically altered through cumulation of 'secondary' or incremental institutional changes such as modifications in contractual relations or shifts in the boundaries between market and non-market activities. (ibid.)

The demand for institutional innovation may stem from a reassignment of property rights, new contractual arrangements or "more efficient market institutions" (ibid.) or, in cases "where externalities are involved, substantial political resources may have to brought to bear to organize non-market institutions in order to provide for the supply of public goods" (ibid.). Supply, on the other hand, is determined by "growing disequilibria in resource allocation due to institutional constraints generated by economic growth [that] create opportunities for political entrepreneurs or leaders to organize collective action to bring about institutional change" (ibid.). The supply of institutional innovation depends on the "struggles among various vested interested groups" (ibid.) as well as the ensuing "cost of

[120] *Open-access*, as described in the article by Hardin (1968) shall not be viewed as a kind of ownership, for open-access is a situation underlying no ownership structure at all.

achieving social consensus (or of suppressing opposition)" (ibid.). The level of costs for that an institutional change be accepted in society depends on the power structure among these interests groups as well as on the underlying cultural tradition and ideology, e.g. nationalism, "that make certain institutional arrangements more easily acceptable than others" (ibid.). Institutional innovation is supplied, according to the authors,

> ...if the expected return from the innovation that accrues to the political entrepreneurs exceeds the marginal cost of mobilising the resources necessary to introduce the innovation. To the extent that the private return to the political entrepreneurs is different from the social return, the institutional innovation will not be supplied at a socially optimum. (ibid.: 213)

In contrast to the neoclassical agenda, the model assumes to explain institutional change as endogenous; but seeing that the divergence of economic returns emanates "from changes in resource endowments and technical change (...) it is a response to exogenous disequilibria in market processes" (Bromley 1989b: 23). As is outlined by the authors,

> ...disequilibria between the marginal returns and the marginal costs of factor inputs occurred as a result of factor endowments and technical change. Institutional change, therefore, was directed toward the establishment of a new equilibrium in factor markets. (Ruttan & Hayami 1984: 209)

With view to the model's supply side, a change toward new equilibria e.g. in factor prices, as a result of legal changes that led to new income opportunities, first, cannot only be viewed as constraints resulting from institutions but come, on the contrary, into existence as a result of collective action that not only restraints but liberates and expands individual action (Bromley 1989: 27). It is thus a change in institutional arrangements that leads to modifications vis-à-vis new income opportunities or resource endowments. Secondly, the model does not account for the ensuing divergence between private and social returns that might stem from institutional innovation not only neglects certain relevant costs that may be shifted to the public, but ignores "the distributional implications of new institutional arrangements created by sheer imbalance of economic power" (ibid. 1989: 25). In addition, the model's driving demand for "more efficient market institutions" (Ruttan & Hayami 1984: 205) thus deploys "concepts of – and judgments about – efficiency and optimality [that] are dependent upon the status quo institutional structure which defines what is a cost and for whom" (Bromley 1989b: 40). In particular, the author highlights that...

> [b]y endogenizing institutional innovation in this manner one is left precisely where conventional welfare economics leaves us – able to comment on changes that seem to be efficient, but unable to comment on the important distributional issues that are at the core of institutional innovation. (ibid. 1989: 25)

On the other hand, as is emphasized by Hagedorn (1991: 64), "[w]hat is regarded as 'efficient' is defined by the existing institutions themselves and thus cannot bring about their change". Hence, any "modifications in contractual relations or shifts in the boundaries between market and non-market activities" (Davis & North 1970: 9) simply represent "a change in the ways in which individuals interact" (Bromley 1989b: 22). It is the exact meaning of institutions, shortly defined as the "rules and conventions that define individual behaviour" (ibid. 1989: 22); for they likewise "define and protect income streams (property rights) it is impossible to have new technology introduced without congenial and appropriate institutional arrangements" (ibid.: 27). Seeing that changes in market and state-led activities depend on "the complementary development of voluntary organizations and norms" (North 1990b: 64), Ruttan and Hayami's (1984) model is not clear-cut in distinguishing institutions from organizations, i.e. purposely created social systems that are comprised by a set of rules that both give meaning to the outside world and define the internal structure (see chapter 2.2.2). The next model on endogenous institutional change fills that niche and draws explicitly attention to the interplay of institutions and organization to understand institutional change.

3.3.2.3 Endogenous institutional change model

The endogenous institutional change model developed by North (1990) and his colleagues (see e.g. Davis & North 1970; North & Thomas 1977) is a modified rational choice approach to institutions, placing human cooperation in the center of thought (Bromley & Yao 2007: 2). According to these scholars, institutions affect the performance of the economy by their effects on the costs of exchange and production; institutional evolution is thusly the key to make the past intelligible (North 1990b: vii). In North's (et al.) model of institutional change special emphasis is drawn "on the continuous interaction between institutions and organizations in the economic setting of scarcity and hence competition" (North 1993: 1). *Institutions* are defined as "the humanly devised constraints that structure political, economic and social interaction" (North 1993: 97) who consist of formal rules, informal constraints and their respective enforcement characteristics.[121] In line with the classification presented above (see chapter 2.2), institutions may be of informal nature (as sanctions, taboos, conventions or other codes of behavior), so-called *informal constraints* (see chapter 2.2.1), or appear in the form of legal rules (e.g. property rights, laws or constitutions), so-called *formal rules* (see chapter 2.2.2) that both – by being designed (e.g. the U.S.' constitution) or have

[121] As already mentioned above, the view on informal rules as usually constraining choice by North (1990) is critically outlined by Bromley (1989b: 27) who stresses that institutions and institutional change are outcomes of collective action that not only restrain individual action but may also liberate and expand it. Likewise, Hodgson who regards institutions as "*social rule-systems*" (2006: 13; italics in the original) emphasizes the role of institutions as expanding "thought, expectation, and action" (ibid.: 2).

evolved over time (e.g. common law) – create incentive structures that ultimately shape the way of economic exchange and overall economic performance (North 1991: 97).[122] More explicitly, these "constraints define (together with the standard constraints of economics) the opportunity set in the economy" (North 1993: 1). It is particularly informal constraints that provide a framework to organize activities and "simplify life" (Colson 1974: 52), seeing that those "not only connect the past with the present and the future, but provide us with a key to explaining the path of historical change" (North 1990: 6). *Formal rules* comprise political (including judicial) rules, economic rules and contracts which form a hierarchy of rules "from constitutions, to statute and common laws, to specific bylaws, and finally to individual contracts that defines constraints, from general rules to particular specifications" (ibid. 1990: 47). To alter rules situated at the upper end of the hierarchy is therefore costlier than those at the lower end of the range, e.g. "…constitutions are designed to be more costly to alter than statute laws, just as statute law is more costly to alter than individual contracts" (ibid. 1990: 47). *Political rules* define the polity's hierarchy, its basic structure of decision-making and "explicit characteristics of agenda control" (ibid.).[123] *Economic rules*, in specific, "define property rights, that is the bundle of rights over the use and the income to be derived from property and the ability to alienate an asset or a resource" (ibid.). In line with the property-rights school (see chapter 3.2.1) North's (et al.) model on institutional change assumes that property rights are created "when it becomes worthwhile to incur the costs of devising such rights" (ibid.: 51), while the evolution of property rights is a "a simple function of changes in economic cost and benefits" (ibid.). These rules serve to facilitate exchange in the political or economic domain while a change "in one will induce changes in the other" (ibid.: 48). It is, however, the "structure of rights (and the character of their enforcement) [that] defines the existing wealth-maximizing opportunities of the players, which can be realized by forming either economic or political exchanges" (ibid. 47).

[122] Hodgson (2006: 11–13) criticizes North's classification of institutions into formal rules and informal constraints, for raising confusion on the meaning of "formal" rules being presumably "legal" and referring to informal rules as ostensibly non-legal (ibid.: 2006: 11).

[123] To present shortly, the polity is modeled at the outset in a simplified way, i.e. illustrated by a ruler and a number of constituent with varying bargaining power (Buchanan & Tullock 1962); it is then expanded by the concept of a representative body who reflects the constituent groups' interests and their role in bargaining with the ruler (whose aim is to receive revenues from the constituent groupings); the increasing involvement of the populace in the process of political decision-making leads, on the one hand, to an "evolution of polities from single absolute rulers to democratic governments" (North 1990: 51) and, on the other hand, supports the development of "third-party enforcement of contracts with an independent judiciary" (ibid. 1990: 51). With view to the latter, *third-party enforcement* refers to "the development of the state as a coercive force able to monitor property rights and enforce contracts effectively" (ibid. 1990: 59; "effectively" carries the meaning of a state who undertook (from an ex-post point of view) successful growth-supporting measures.

Contracts are defined as to "contain the provisions specific to a particular agreement in exchange" (ibid.: 47) and "will reflect the incentive-disincentive structure embedded in the property rights structure (and the enforcement characteristics)" (ibid.: 52). Moreover,

> [c]ontracts provide not only an explicit framework within which to derive empirical evidence about the forms of organization (and hence are the basic empirical source for testing hypotheses about organization), but also clues with respect to the way by which the parties to an exchange will structure more complex forms of organization (…) that extend in a continuum from straightforward market exchange to vertically integrated exchange. (ibid. 53; cf. (Williamson 1975)

Economic performance depends on low costs of contracting and enforcing those contracts, one of the key problems of institutional change (North 1990: 54–55). Thus, North (1990) stresses that contracts are the most self-enforcing where the "parties to exchange have a great deal of knowledge about each other and are involved in repeat dealings (…). Under these conditions, it simply pays to live up to agreements" (ibid.: 55). In contrast, a major problem arises as a result of the inevitable incompleteness of contracts and their subsequent measurement and *enforcement* problems. From the perspective of wealth-maximizing individuals, in cases "where there are high costs of measurement and no form of enforcement is possible, the gains from cheating and reneging exceed the gains from cooperative behavior" (ibid.). Thus, as institutions together with the technology applied make up total transaction costs (ibid.: 61), the costs of contracting and enforcement depend especially on the property rights structure, the performance of the court and judiciary, and "the complementary development of voluntary organizations and norms" (ibid.: 64; see chapter 2.2.4). For example, the size of an asset's discount rate "will be greater to the degree that the institutional structure allows third parties to (…) affect the value of the property" (ibid.: 63).

Organizations, on the other hand, are a specific sort of institution, comprised by groups of individuals that are "bound by some common purpose to achieve objectives" (ibid. 1990: 4).[124] They provide, as do institutions *per se*, a framework for human interaction comprised by political entities (as parliaments, political parties, a regulating agency or a city council), economic units (as family farms, firms, trade unions or cooperatives), social groupings (as clubs, churches or sports associations) and educational bodies (as kindergartens, schools or vocational training centers) (ibid.: 5). The specific formation of these going concerns (see chapter 2.2.2) – "[b]oth what organizations come into existence and how they evolve"

---

[124] Bromley (1989b: 27) sees "the treatment of institutions as both rules of organizations, and as the organizations themselves" as one of the shortcomings in North's model of institutional change. Likewise does Hodgson (2006: 9–10) who criticizes North's definition of organizations that are implicitly treated as unitary players, neglecting "the potential conflict within the organization" (ibid. 2006: 9) and the organization's internal embedded rule structure that define "rules of communication, membership, or sovereignty" (ibid.: 10).

(ibid.: 5) – is a result of the *institutional framework.* [125] Organizations, in turn, "influence how the institutional framework evolves" (ibid); they are formed…

> …with purposive intent in consequence of the opportunity set resulting from the existing set of constraints (institutional ones as well as the traditional ones of economic theory) and in the course of attempts to accomplish their objectives are a major agent of institutional change. (ibid.: 5)

The purposeful creation of organizations by (political or economic) entrepreneurs is based on efforts "to maximize wealth, income, or other objectives defined by the opportunities afforded by the institutional structure of the society" (ibid.: 73). However, due to the dominant role of path dependendent historical developments, "…the institutional framework of a polity (and economy) is characterized by increasing returns so that incremental change is heavily weighted in favor of policies consistent with the basic institutional framework (North 1990a: 364). The subsequent rise of organizations hence "reflect the opportunities available in that institutional setting" (ibid. 1990a: 364). It is exactly "[t]he symbiotic relationship between institutions and the consequent organizations [which] shape the direction of political/economic exchange" (ibid.: 365). North hence emphasizes that "the overall direction of the polity or economy is difficult to referse" (ibid.). As institutions shape human behavior they are modeled as "the rules of the game in a society" (ibid.: 3); to modeling organizations and affiliated individuals, i.e. their members, comprise its internal governance structure and forms of their members' interaction, namely the "combination of skills, strategy, and coordination" (North 1990b: 4–5), and "how learning by doing will determine the organization's success over time" (ibid.: 5; cf. (Nelson & Winter 1982).[126] The latter is thus endogenously defined and depends on individuals' particular ability to acquire *tacit knowledge* (Polanyi 1967), i.e. knowledge and coordination skills mostly attained by repeated interaction in practice; it is distinct from personally transmitted knowledge, so-called *communicable knowledge* (North 1990b: 74). The incentives to learn and to acquire skills and knowledge stem from "the structure of the monetary rewards and punishments, but also by a society's tolerance of its development…" (ibid. 1990: 75). The perception of the latter is a result of "the way knowledge develops [which] shapes our perceptions of the world around us and

---

[125] The latter is specified by the institutional environment which is "a set of the fundamental political, social, and legal ground rules that govern economic and political activity (rules governing elections, property rights, and the rights of contract are examples of these ground rules)" (Davis & North 1970: 133). A particular institutional arrangement then describes "an arrangement between economic units that governs the ways in which these units can cooperate or compete" (ibid. 1970: 133).

[126] Though institutions in general are said to facilitate exchange, North takes into account that institutions lower but may also increase transaction costs, e.g. due to rules that restrict entry, raise information costs "or make property rights less secure" (North 1990b: 63).

in turn those perceptions shape the search for knowledge" (ibid.: 76). The key of an organization's success relates, however, to its specific context, as it…

> …not only shapes the internal organization and determine the extent of vertical integration and governance structure, but also determine the pliable margins that offer the greatest promise in maximizing the organization's objectives. (ibid.: 77)

Besides of its institutional environment, further influential features for an organization's success are "competition, decentralized decision-making, and well-specified contracts of property rights as well as bankruptcy laws" (ibid. 1990: 81). Thus, distinct set of rules will hence "produce different incentives for tacit knowledge" (ibid. 81). The institutional framework affects the learning process of acquiring skills and knowledge, while the very course of learning is "the decisive factor for the long-run development of that society" (ibid.: 78).[127] The maximization efforts of economic organizations affects institutional change, first, by the ensuing "demand in knowledge of all kinds" (ibid.); second, by the resultant interaction between the organization's economic activities, "the stock of knowledge, and the institutional framework (…)" (ibid.); and, third, by by-producing alterations of the underlying informal constraints (ibid.).

The institutional framework moreover provides the incentive structure of economic activity, i.e. its payoffs. The organization's entrepreneurs "induce institutional change as they perceive new or altered opportunities" (ibid. 1993: 1), for instance

> …by altering the rules (directly in the case of political bodies; indirectly by economic or social organizations pressing political organizations); or by altering, deliberately and sometimes accidentally, the kinds and effectiveness of enforcement of rules or the effectiveness of sanctions and other means of informal constraints enforcement. (ibid.: 1)

The course of institutional change is thus shaped, first, by the interdependence of institutions and organizations and the incentive structure of the institutional framework, and, second, by the "feedback process by which human beings perceive and react to changes in the opportunity set" (ibid. 1990: 7).[128]

---

[127] The "systematic investment in skills and knowledge and their application to an economy" (ibid.: 80) is measured (not by Pareto conditions defining the *allocative efficiency* of the neoclassic economics school's but) by *adaptive efficiency* which "is concerned with the kind of rules that shape the way an economy evolves through time" (ibid.).

[128] In contrast to North's endogenously derived assumptions of institutional change, Bromley and Yao (2007) view the "essential interplay between institutions as both exogenous and endogenous rule structures" (Bromley & Yao 2007: 3): Endogenous rule structures emerge by the interplay of individuals and thus become institutions (in contrast to predictable patterns) (ibid. 2007: 3–4).[128] On the other hand, exogenous patterns emerge in both so-called democratic nation states as well as authoritative political systems with a recognized and enforced constitutional order defining both constitutional rules as well as organizational or "instrumental rules of the economy" (ibid.: 5). Likewise Ostrom (2005) defines exogenous

The North's (et al.) model is hence critical toward behavioral postulates of the neoclassical school inhibit in assumptions on individuals' profit maximization and stable preferences; instead, North (1990) argues that incremental change derive

> ...from the perceptions of the entrepreneurs in political and economic organizations that they could do better by altering the existing institutional framework at some margin. But the perceptions crucially depend on both the information that the entrepreneurs receive and the way they process that information. (ibid.: 8)

The author hence relates to "two particular aspects of human behavior: (1) motivation and (2) deciphering the environment" (ibid: 20). (1) The author regards the idea to model complex individual behavior by utility functions of ostensible wealth-maximizing individuals critical and hence supports, quite the reverse, the believe in peoples' decision-making based on "ideology, altruism and self-imposed standards of conduct (...) frequently playing a major role in the choices individuals make" (ibid: 22). (2) Instead of the school's assumption on complete information and stable preferences in individual's decision-making, North's model focuses on individuals' perception as a cognitive process by taking account of individuals' restrained processing capabilities due to people's own mental models.[129]

An alteration might be a result of "changes in rules, in informal constraints, and in kinds and effectiveness of enforcement" (ibid.: 6). As the institutional framework is stabilized "by a complex set of constraints that include formal rules nested in a hierarchy" (ibid.: 83), a change thereof demands vesting substantial resources, "where each level is more costly to change than the previous one" (ibid.). The set also comprises informal constraints which extend, elaborate and qualify rules that prove "tenacious survival because they have become part of habitual behavior (ibid.). Thus, while it is possible to change legal rules, as laws, from one day to the other by political consent, changes in informal rules, as exhibited in custom or traditions, "are much more impervious to deliberate policies" (North 1990b: 6).[130] As a result, "changes at the margin may be so slow and glacial in character that we have to stand back (...) to perceive them..." (ibid.). Hence, institutional change can be observed by "this interplay between institutions (rules) that are imposed on lower level entities in an economy, and the endogenous response emerging from these lower level going concerns" (Bromley & Yao 2007: 8). Seeing that "imposed institutional arrangements are the conscious and purposeful in-

variables in her analysis of institutional change, namely rules, the attributes of the community or the biophysical and material conditions (Ostrom 2005: 14).

[129] The latter is thus based on features depicted by Herbert Simon's concept on bounded rationality (Simon 1972, 1983).

[130] The differences of formal and informal rules are specified in chapter 3.1.

struments of national (and provincial) economic policy" (ibid. 2007: 8) those inhibit an exogenous element. E.g. rules stipulating that "[a]gricultural land must not be privatized, but other forms of contracts among farmers and local officials may be worked out" (ibid.: 8) show "that imposed institutions do not simply constrain individual and group behavior – they also liberate behavior" (ibid.). Farmers find new ways of interaction and hence "we see endogenous institutional change at work — and correlated evolved behavioral patterns" (ibid.).

These changes are typically *incremental* for revolutions occur rarely (Tullock 1977; Skocpol 1994). In case of the latter, North refers to *discontinuous institutional change* that leads to radical changes in formal rules; *continuous incremental change*, in contrast, is led by "continuous marginal adjustments" (North 1990: 101), provided by an "institutional context that make possible new bargains and compromises" (ibid. 1990: 89). In cases of discontinuous change North underlines though that "it is seldom as discontinuous as it appears on the surface (…). It is seldom so discontinuous partly because coalitions essential for the success of revolutions tend to have a short afterlife" (ibid.: 90). Though even if

> …a wholesale change in the formal rules may take place, at the same time there will be many informal constraints that have great survival tenacity because they still resolve basic exchange problems among the participants, be they social, political or economic. (ibid.: 91)

A change of enforcement mechanisms may provide new profitable opportunities to some organizational entrepreneurs "with new avenues of profitable exploitation that in turn shift the direction of institutional change" (ibid.: 88). The two major sources leading an institutional framework to alter are changes in relative prices and altered preferences (ibid.: 84). The most important source though are fundamental changes in prices, particularly "changes in the ratio of factor prices (i.e. changes in the ratio of land to labor, labor to capital, or capital to land), the cost of information, and changes in technology (including significantly and importantly, military technology)" (ibid.). Moreover, in case

> …that there are large payoffs to influencing the rules and their enforcement, it will pay to create intermediary organizations (trade associations, lobbying groups, political action committees) between economic organizations and political bodies to realize the potential gains of political change. (ibid.)

In some cases, relative price changes stem from exogenous sources, e.g. as a consequence of plagues, "but most will be endogenous, reflecting the ongoing maximization efforts of entrepreneurs (political, economic, and military) that will alter prices and in consequence induce institutional change" (ibid.). E.g. efforts are undertaken to renegotiate contracts in the political or economic domain as a result of a change in bargaining power (ibid.). In case of technological changes, the concept is based on a path dependent evolution of technology where incremental changes toward one technological path, "once begun on a particular track, may lead one technological solution to win out over another, even when, ultimately, this technological path may be less efficient than the abandoned alternative would have been" (ibid.: 93). The dominant role of one technological solution, in case of competing technologies, might be the result of "increasing returns [which] imply a single winner over time. Or, simply, some small event may give one technology an advantage over the other" (ibid.: 94). Arthur (1988) identifies the following patterns that reinforce a dynamic toward one economic actor gaining a monopolistic position over others: (1) Large setup or fixed costs "which give the advantage of falling unit costs as output increases" (North 1990: 94); (2) learning effects that either lead to an increase in quality or the lowering of costs; (3) coordination effects among those employed in the same field who take advantage of cooperation with each other; and (4) adaptive expectations as a result of "increased prevalence on the market [that] enhances beliefs of further prevalence" (ibid. 1990: 94). While the final outcome is characterized as (i) indeterminate, the outcome may, according to Arthur (1989) result in (ii) possible inefficiencies "because of bad luck in gaining adherence" (North 1990: 94), (iii) a lock-in, i.e. a situation which is difficult to exit from once reached (ibid. 1990: 94); or (iv) path dependence, "the consequence of small events and chance circumstances [which] can determine solutions that, once they prevail, lead one to a particular path" (ibid.: 94). The competition between technological solutions though is only indirectly; the competition is directly led by the organizations employing these technologies (ibid.): "The distinction is important because the outcome may reflect differing organizational abilities (tacit knowledge of the entrepreneurs) as much as specific aspects of the competing technologies" (ibid.). Moreover, the evolution of technology is treated endogenously in this model, in contrast to the neoclassical economics' notion, considering technical change not being part of human organization (ibid.: 132).

Changes in tastes and preferences are usually a by-product of price changes. The latter is at one point perceived by one or both sides of the contract, "whether it is political or economic, (…) that either or both could do better with an altered agreement…" (ibid.: 86), and try to negotiate it. In case that a renegotiation does tackle rules nested in a hierarchy of rules may either demand a restructuring of rules on a higher level, or may violate some norm of behavior (ibid.). The latter though requires an institutional framework "that allows people to express their views at little cost to themselves" (ibid.: 85). If that is the case, a perceived improvement by a change in behavior may lead to the formers' gradual erosion being replaced by another, even though in the course of time, also the latter "may be changed or simply be ignored and unenforced. However, this rather simplified story gets more complicated in many ways – by agenda power, by the free-rider problem, or by the tenacity of norms of behavior" (ibid.: 86).

North thus emphasizes the key role of informal constraints that "modify, supplement, or extend formal rules" (ibid.: 87) or their enforcement mechanisms. In cases of informal constraints that are no longer viewed to "meet the needs of newly evolved bargaining structures", formal rules might be "developed deliberately to overrule and supersede existing informal constraints" (ibid.: 88). However, the result of institutional change differs, especially with view to distinct historical developments among people confronted with "different problems with different resource endowments, different human capabilities, and in different climates" (ibid.: 92). The outcome hence may not necessarily be socially productive, "because the institutional framework frequently has pervasive incentives" (ibid.): "The resultant organizations will evolve to take advantage of the opportunities defined by that framework, but as in the case of technology, there is no implication that the skills acquired will result in increased social efficiency" (ibid.: 95). Thus, gradual institutional change is the outcome of alterations of incentives and constraints at the margin that define opportunities, "and the margins affected will be those where the immediate issues require solution and the solution will be determined by the relative bargaining power of the participants" (ibid.: 101). This, in turn, has consequences and defines the various paths of institutional change, seeing that

> ...the bargaining power of groups in one society will clearly differ from that in another, the marginal adjustments in each society will typically be different as well. Moreover, with different past histories and incomplete feedback on the consequences, the actors will have different subjective models and therefore make different policy choices. (ibid.)

With view to the fact that institutional change is advanced by economic and political entrepreneurs "who attempt to maximize at those margins that appear to offer the most profitable (short-run) alternatives" (ibid.: 100), the adjustments may "result in the pursuit of persistently inefficient activities" (ibid.) for biased by a higher discount-rate. With view to the different effects of applying blue-print approaches, it is hence no surprise that although

> ...the rules are the same, the enforcement mechanisms, the way enforcement occurs, the norms of behavior, and the subjective models of the actors are not. Hence, both the real incentive structures and the perceived consequences of policies will differ as well. (ibid.)

The implications for analyzing institutional change is thus to "relate institutions to incentives to choices to outcomes" (ibid.: 134) by focusing on "what institutional characteristics have shaped performance" (ibid.). The organizational forms chosen in exchange structures consequently reflect "unequal access to resources, capital, and information" (ibid.). A systematic study of the underlying institutional arrangements therefore enquires the sources of institutional change: Is the institutional structure, first, "imposed from without or is it endogenously determined or is it some combination of both?" (ibid.). Second, by obtaining empirical information "on transaction and information costs" (ibid.: 135) it is possible to "trace the institutional origins of such costs" (ibid.) and hence to understand an economy's performance.

3.3.3 Institutional change led by (and controversies on) foreign aid

According to Radelet (2006: 4) foreign aid, or foreign assistance, is typically defined by the Development Assistance Committee (DAC) of the Organization for Economic Cooperation and Development (OECD)

> ...as financial flows, technical assistance, and commodities that are (1) designed to promote economic development and welfare as their main objective (thus excluding aid for military or other non-development purposes); and (2) are provided as either grants or subsidized loans. (ibid. 2006: 4)

Financial means provided by grants and subsidized loans comprise *concessional financing*, while "loans that carry market or near-market terms (and therefore are not foreign aid) are *non-concessional* financing" (ibid.: 4).[131]

Three classifications are made in accordance to the per capita income of the recipient country, namely *Official development assistance, Official assistance* and *Private voluntary assistance*. The first, *Official development assistance* (ODA) is comprised by "aid provided by donor governments to low- and middleincome countries" (ibid.). *Official assistance* (OA) is a form of "aid provided by governments to richer countries with per capita incomes higher than approximately $9,0005 (e.g., Bahamas, Cyprus, Israel and Singapore) and to countries that were formerly part of the Soviet Union or its satellites" (ibid.). The alternative *Private voluntary assistance* is based on "grants from non-government organizations, religious groups, charities, foundations, and private companies (ibid.).

Two further categories relate to the contributor(s) of assistance, i.e. whether provided as bilateral or multilateral assistance. According to Radelet (2006: 5), "[h]istorically most aid has been given as bilateral assistance directly from one country to another". If resources are pooled which are provided by many donors, multilateral assistance is given indirectly through multilateral organizations, e.g. through the many different United Nations agencies, as the United Nations Development Programme (UNDP), the World Bank or "the African, Asian, and Inter-American Development Banks" (ibid. 2006: 5).

The U.S. has, in terms of total dollars, "consistently been the world's largest donor (except in the mid-1990s when Japan briefly topped the list)" (ibid.: 5). In relative terms though, i.e. aid measured as share of a contributor's income, the U.S. "is one of the smallest donors by this measure at about 0.17 percent of U.S. income in 2004, just over half of the 1970 level of 0.32 percent and less than one-third of the U.S. average during the 1960s" (ibid.). In contrast, "the most generous donors are Norway, Denmark, Luxembourg, the Netherlands, and Sweden, each of which provided between 0.79–0.92 percent of GDP in 2004. Saudi Arabia provided aid equivalent to about 0.69 percent of its income" (ibid.). But although the

---

[131] For an overview of the exact conditions applied for loans to carry market or near-market terms, see Radelet (2006: 4).

international donor community has "pledged since the 1960s to devote 0.7 percent of their income as aid […] only a handful of small donors have achieved this level of aid" (ibid.).

The reason though to provide foreign assistance is according to Alesina & Dollar (2000: 41–42) less movitated by a country's poverty level, degree of democracy or political institutions than led by political-strategic considerations. E.g. the motivation led by…

> …the "big three" donors – U.S., Japan, and France – has a different distortion: the U.S. has targeted about one-third of its total assistance to Egypt and Israel; France has given overwhelmingly to its former colonies; and Japan's aid is highly correlated with UN voting patterns (countries that vote in tandem with Japan receive more assistance). These countries' aid allocations may be very effective at promoting strategic interests, but the result is that bilateral aid has only a weak association with poverty, democracy, and good policy (ibid. 2000: 55).

Moreover, the authors show that …

> …[a]fter controlling for its special interest in Egypt and Israel, U.S. aid is targeted to poverty, democracy, and openness. The Nordic countries have a similar pattern except that they do not have the same sharp focus on the Middle East. French assistance, on the other hand, has little relationship to poverty or democracy even after controlling for their strategic interests in former colonies and UN friends. The same conclusion holds for Japan, with the caveat that its strategic alliance may be built around investment and trade relationships, more than former colonial ties (ibid.: 55–56).

The effects and effectiveness of foreign aid are still under discussion.[132] According to Easterly (2003: 26), widespread theoretical discussions on the reasons for poverty and growth started from the 1950s, but empirical data was either not available, nor sufficiently convincing.[133] Although Burnside & Dollar (2000) present

---

132 For a literature overview on the effectiveness of aid, see e.g. Hansen & Tarp (2000).

133 Lewis (1954) and Rostow (1971) pronounced capital, growth and modernization theories during the Cold War, justify colonialism and imperialism and aimed to replace traditional "underdeveloped" (and hence endogenously growth hampering) norms, values and practices by modern, progressive (exogenously induced growth supporting) norms, values and behaviors. With a rising number of states becoming independent, as e.g. India and Egypt, so-called dependencia theories emerged that center on domination and exploitation in international trade relations. Foremost linked to the Argentinian and German economists Raùl Prebisch and Wolfang Singer their work focus on countries' terms of trade to explain the state of its industrialization or underdevelopment, according to the sale of industrialized manufactured goods in the first place and trade based on primary goods in the latter, see e.g. Harvey et al. (2010); other representatives of this stream of thought comprise Gunder Frank (1966) or Cardoso & Faletto (1979). With statements as "[w]hite men saving brown women from brown men" (Spivak 1988: 293) feminist scholars emphasize women's subordination in the discourse of gender and development, see e.g. Spivak (1988), Mohanty (1991) or Wieringa (1994).

a positive correlation between growth and foreign aid, e.g. Boone (1996), in contrast, shows that foreign assistance favors less the poor but the political elite who becomes the main recipient of aid flows. Thus, Radelet (2006: 39) concludes, "[h]ow to achieve a beneficial aggregate impact of foreign aid remains a puzzle". Morover, the author underlines that the notion "of aggregating all this diversity into a "developing world" that will "take off" with foreign aid is a heroic simplification" (ibid. 2006: 39). However, Ostrom et al. (2014: 117) highlight that…

> [s]uccessful development aid generates the appropriate incentives so that the time, skill, knowledge, and effort of multiple individuals create jointly valued outcomes. These incentives come from the institutions – the rules of the games of life and productive coexistence – that developmemt aid helps to create or modify.

The authors stress that – in contrast to the general understanding of aid agencies who operate "as external mechanisms that infuse the capital and expertise needed to relieve the pathologies of poverty and transform societies" (ibid. 2014: 117) – "concepts as collective action and collective-action problems are central to understanding the challenge of development as well as core environmental problems" (ibid.: 118).[134] As a consequence, the authors suggest to direct "…more attention to institutional analysis during all stages of the development cooperation process, [so that] performance and sustainability are likely to improve" (ibid.).

3.4 Analyzing institutional change: The Institutions of Sustainability (IoS) Framework

3.4.1 Analytical categories and the unit of analysis

The analytical framework used for the study is adapted from framework "Institutions of Sustainability" (IoS) (Hagedorn 2008; Hagedorn, Arzt & Peters 2002; see figure 3-4 below), "which focuses on how to regularise human action that leads to transactions affecting the relationship between natural and social systems" (Hagedorn 2008: 359). The framework therefore allows for an institutional analysis of the (changing nature of) interactions between human and natural systems in the context of agriculture or any other form of resource management, and thus equally qualifies to institutional changes vis-à-vis land governance in transition economies (ibid. 2008: 358–359).

[134] A *collective-action situation* is understood "as a situation that occurs when two or more individuals come together to produce something of value, when it would be difficult to produce it alone" (Ostrom et al. 2014: 118); *collective-action problems* occur "when a lack of motivation and/or missing or asymmetric information generates incentives that prevent individuals from satisfactorily resolving a collective-action situation" (ibid. 2014.: 118).

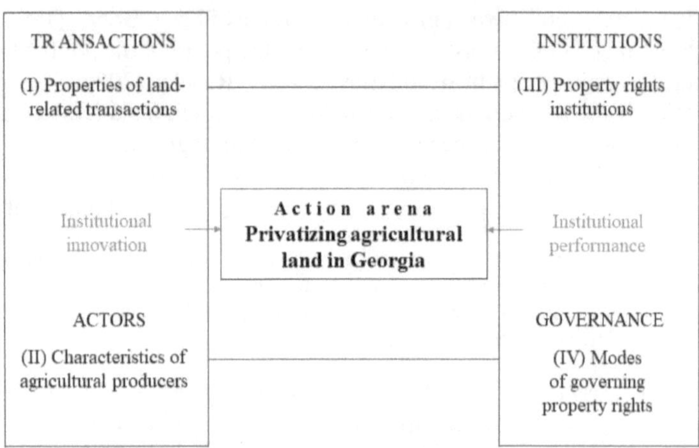

Figure 3-4: Institutions of Sustainability (IoS) Framework (Hagedorn 2008; Hagedorn, Arzt & Peters 2002)

The IoS framework is comprised by four components, namely the *characteristics of the actors involved,* the *properties of the actors' transactions* that link social and natural systems, as well as *institutions* and *governance*. The unit of analysis is comprised by t*ransactions* which "present economically relevant processes by which goods and services, resources and amenities, and damages and nuisances are allocated" (ibid.: 360).[135] At the focal point are nature-related economic activities and their institutional arrangements, e.g. farming or cutting of trees, which generate transactions. Seeing that resources are scarce, transactions produce interdependencies among actors (ibid.: 361). *Actors* are defined as those "who are able to *consciously select* what action they want to take" (Hagedorn 2008). Accordingly, it is the "…identity of the people engaged in a transaction [which] is a major determinant of the institutional mode of transaction" (Ben-Porath 1980: 1). The type of transaction typically depends on impersonal relationships, namely

> [t]he degree to which identity dominates or is subsumed under the impersonal dimensions of specialization [which] shapes the type of transaction or contract. The family is the locale of transactions in which identity dominates; however, identity is also important in much of what we consider the "market"… (Ben-Porath 1980: 1)

[135] The term was initially shaped by Commons (1931) to refer to economic activities which relate not specifically to "the "exchange of commodities," in the physical sense of "delivery," they are the alienation and acquisition, between individuals, of the *rights* of future ownership of physical things…" (Commons 1934: 58); italics in the original). The term was further applied by O. Williamson (1985) to understand organizational forms within firms that are costs saving.

The *action arena* is defined as a "situation in which a particular type of action occurs" (Ostrom 2005: 32), and consists of an "action situation, and individuals and groups who are routinely involved in the situation (actors)" (Polski & Ostrom 1999: 6). Moreover, "[a]ction arenas exist in the home; in the neighborhood; in local, regional, national, and international councils; in firms and markets; and in the interactions among all of these arenas with others" (Ostrom 2005: 12). This interaction, in turn, causes the actors' interconnectedness and mutual dependence (Hagedorn 2008: 361). *Institutions* are, as presented above (see chapter 3.1), broadly understood as "collective action in restraint, liberation and expansion of individual action" (Commons 1934: 73). Seeing that institutions are defined as "systems of established and embedded social rules that structure social interaction" (Eggertsson 1990: 18) they reduce uncertainty "by establishing a stable (but not necessarily efficient) structure to human interaction" (North 2006: 6).[136] *Governance* defines organizational modes to make institutions to become effective, e.g. via contracts, markets or bureaucracies (Hagedorn 2008: 360; see chapter 3.2). The transfer of entitlements, regularized by institutions and governance structures, is based on so-called *institutionalized transactions,* which emphasize its social dimension (Hagedorn 2008: 363). Both institutions and governance systems are aligned and shape the *institutional performance* defined as "the situation in which a particular type of action occurs" (Ostrom 2005: 32). Seeing that the institutional environment is constantly evolving (North 1990b: 6), the properties of the transactions together with the properties of the actors lead to *institutional innovation* (Hagedorn 2008: 358).

As is emphasized by Thiel (2006: 30) though, the IoS-framework does not take into account the interdependencies of physical and institutional time lags: The former refers to the (contingently aggravated) problem manifestation in natural systems, while the latter applies to potential changes of the governance structure that might impact on future resource use. Thus, to evaluate institutional arrangements requires that "the development trend of the activities and transactions governed should be characterised as well as the longevity of the environmental and coordination problem they produce" (ibid. 2006: 30). The analysis thus firstly takes an ex-ante viewpoint on institutional change, before analyzing its outcomes (ex-post) (Hagedorn 2008: 361). It shall thereby be possible to "decompose the process into a series of stylised steps to illustrate how and why physical transactions become institutionalsed transactions" (ibid. 2008: 361).

[136] In his study North (1990: 92) indicates conditions as *efficient* "where the existing set of constraints will produce economic growth".

3.4.2 Properties of nature- and ecosystem-related transactions

The *Properties of nature- and ecosystem-related transactions* are influenced by the physical characteristics of a resource, and are equally affected by the interdependent relations of the actors involved (Hagedorn 2008: 361–362).[137] Paavola and Adger (2005: 364) underscore, "the institutional approach helps us to examine how the attributes of environmental resources and their users create interdependence and conflicts". Ecosystem attributes relate to the *stock*, e.g. land as a renewable resource, as well as to its *flows* as a result of e.g. land use, i.e. harvesting and pasturing (Hagedorn 2008: 375; cf. Low et al. 1999). Nature-based features of transactions refer to *jointness* or *lack of separability* on the one hand, as well as *coherence* and *complexity* on the other; they may involve indirect or direct transfers, "have a spatial dimension, involve time lags, be complicated to reproduce or even be hidden; transactions may be intended or unintended, targeted or non-targeted, predictable or unpredictable" (Hagedorn 2008: 362).

The decision how to regularize a transaction is usually made in accordance with the physical properties of a good and subsequently sorted on a state-market dichotomy – based on the general criteria of *rivalry* and *excludability* – in private and public goods (Hagedorn 2008: 364).[138] However, "[t]his concept moves directly from the attributes of goods to the question of the institutional fit of the market and misses the intermediate step…" (ibid. 2008: 364), namely the "…explicit consideration of the properties of transactions that may be compatible or incompatible with certain institutional and organizational arrangements" (ibid. 2008: 364). Transaction cost economics support the idea that the choice on "[w]hich transaction go where depends on the attributes of transactions on the one hand and the costs and competence of alternative modes of governance on the other" (Williamson 1996: 151). This follows the *discriminating alignment hypothesis* which "holds that transactions, which differ in their attributes, align with governance structures, which differ in their cost and competence, in a discriminating – mainly, transaction cost economizing – way" (ibid. 1996: 153). The "three critical dimensions" to identify the properties of a transaction are (1) *uncertainty*, (2) *frequency* of the transactions to recur, and "(3) the degree to which durable *transaction-specific investments* are incurred" (Williamson 1979: 239),

[137] Transaction cost economics applied to natural systems have several implications, for it comprises according to Hagedorn (2008: 361–362) "resources, goods and services whose transactions involve processes of self-organisation in ecosystems not completely engineered by humans, but often influenced or even disturbed by them". This means that a transfer of e.g. land use rights affects interdependent actors physically (ibid.: 362; Schmid (2004: 69ff.)); e.g. the leasing of some hectares of my land to my neighbor reduces my workload and allows her to use and benefit of that part of land. "The only requirement for an action to be also called a transaction is that the actors involved are affected due to a physical implication" (Hagedorn 2008: 362).

[138] The following lines either refer to "good", "goods", or "goods and services", but relates in fact to both goods and services and shall, however stated, not exclude one or the other.

the latter is further referred to as *asset-specificity*. Although uncertainty is understood to be of critical nature (ibid.), asset-specificity is viewed as the most crucial feature for it relates to transactions in highly specified contractual relationships which may lead to bilateral dependency (or a lack thereof) and incomplete contracting which implies cost-bearing consequences (Williamson 1981, 1971).

(1) The first dimension, *uncertainty*, relates to instances where both parties – situated in a strategically interdependent situation to bargain over future benefits – are confronted with asymmetric information that might be the basis for opportunistic behavior due to contractual ex-ante occurrences of adverse selection or ex-post contractual (moral) hazards (Williamson 2000).[139] In ecosystem-related production systems uncertainty may arise by missing detection of "credence attributes", for example in agri-food industries (cf. Van Huylenbroeck 2003; Hagedorn 2008: 367), or based on imprecise policies and a subsequent lack of accuracy in the implementation of these environmental policies, which at the same time impact on the rise of transaction cost (cf. Vatn 2002). In addition to imperfect foresight, Alchian (1950) detects the inability to solve complex problems as the second source of uncertainty; the imprecise formulation or implementation of policies, as proposed by Vatn (2002: 314) might hence apply to the latter.

(2) *Frequency* applies to the rate of events that may occur as one-time, occasional or recurrent transactions (Williamson 1979). "Frequency is relevant in two respects: reputation effects and setup costs, the net effects of which will vary with the particulars" (Williamson 2005: 14).

(3) *Asset-specificity*, in general, is a fundamental dimension that may play a role with regard to *site*, or with view to *physical* and *human* aspects (Williamson 1981: 555): S*ite specificity* plays a role where the location of stations and its proximity to each other impacts on, e.g., the costs of inventory or transport; with view to ecosystems, this aspect relates to spatial characteristics of a resource and its

[139] According to Williamson (1985), opportunistic behavior is defined as "self-interest seeking with guile" (Williamson 1995: 15) and he outlines, "opportunism refers to the incomplete or distorted disclosure of information, especially to calculated efforts to mislead, distort, disguise, obfuscate, or otherwise confuse" (ibid. 1995: 47), or in other terms, "breach of contracts involving strategic manipulation of information" (Eggertsson 1990: 171). Accordingly, *adverse selection* and *moral hazard* are both examples of contractual incompleteness and thus market failures that are based on asymmetric information between two contractual partners: From an ex-ante point of view, *adverse selection*, as was outlined in the example of Akerlof's (1970) "market for lemons applies to situations where contractual agreements are made on the basis of incentives (selling cars) which lead to the quality's decline (selling lemons, i.e. bad cars) equally to the principle of Gresham's law: "The "bad" cars tend to drive out the good (in much the same way that bad money drives out the good" (Akerlof 1970: 489–490); *moral hazard* applies to contractual ex-post situations in which a behavior for which an insurance is provided, provoke that very behavior. E.g. investors act perilously and take excessive risk if the consequences of their acts are backed, and they are not held liable for their actions (Mishkin 1999).

mobility (Hagedorn 2008: 366); *physical asset specificity* is linked to the decomposability, modularity or independence of the (natural) process in question, which may be comprised by either complex and interconnected, or atomistic and isolated transactions (ibid. 2008: 365–366; see further below); physical attributes of nature-related transactions might additionally stem from "limited standardisability and calculability, dimensions of time and scale, predictability and irreversibility, (…) adaptability and observability, etc." (Hagedorn 2008); *human asset specificity* arises from (symmetrical) experiences in contractual relationships, for "once an investment has been made, buyer and seller are effectively operating in a bilateral (or at least quasi-bilateral) exchange relation for a considerable period thereafter" (Williamson 1981: 555). Thus, besides of asset-specificity being important with regard to investments, Ménard (2004) points to focus on the creation of mutual dependence and alerts: The higher asset-specificity, the higher the risk of opportunistic behavior (Ménard 2004b: 212). "With increasing risk of opportunism, forms of private government develop for coordinating and policing the relationship, moving it away from a contract-based agreement and closer to quasi integration…" (ibid.: 355). This assumption is confirmed by Williamson (1991) who underlines that a "[l]ack of credible commitment on the part of the government poses hazards for durable, immobile investments of all kinds-specialized and unspecialized alike-in the private sector" (Williamson 1991: 289).

In addition, with due consideration of the complexity of nature-related transactions, Hagedorn (2008) introduces "*modularity and decomposability of structures* and *functional interdependence of processes*" (Hagedorn 2008: 372; emphasis in the original) as key attributes in order to conceptualize and "to establish causal relationships between natural system attributes, transaction properties and social constructions for regulating and governing nature–human interactions and actor interdependencies" (Hagedorn 2008: 371). Complete structural modularity eases the analysis of complex systems (Velichkovsky 2005; Simon 1983; Marr 1976). Moreover, it is assumed (in accordance with Coase on the basis of a perfectly competitive market) that complete modularity leads to zero transaction costs for single modules can be adapted to an "optimal" size; non-separability of functional processes, on the other hand, might lead to externalities that can only become internalized by decomposing the system into smaller units (ibid. 372; cf. Marengo et al. 2001: 9–10).[140] The essential point is that even though it is possible to decompose a system into sub-units *structurally*, these components may either work inter-connected or isolated and atomistic, and hence nothing is said about their *functional* interdependency (ibid. 372). Watson and Pollack (2005) underline that "[t]he exact consequence of interactions between modules is dependent on the exact nature of the systems involved…" (Watson & Pollack 2005: 446), and thus argue that "in general, it is not correct to assume that sparsely connected dynam-

[140] The issue of externalities is treated in-depth further below (see chapter 3.1.3).

ical systems have only small effects on one another's dynamical properties". Consequently, transactions are classified "along the gradual continuum between atomistic-isolated transactions and complex-interconnected transactions" (Hagedorn 2008: 373) in order to "enable institutional analysts to identify and order heterogeneous transactions, particularly those found in domains where actors use, manage, degrade and protect natural systems" (ibid. 2008: 374).

3.4.3 *Evaluative criteria by linking action arenas*

Seeing that individuals' interactions do not take place in vacuum, "most of social reality is composed of multiple arenas linked sequentially or simultaneously (Shubik 1986)" (Ostrom 2005: 55). Thus, in addition to a thorough analysis of the effects of exogenous factors on an action situation, "an important development in institutional analysis is the examination of linked arenas" (ibid. 2005: 55). The policy process can therefore be conceptualized "as a complex network of linked action situations, with the outcome from any one node affecting the likely outcomes that will emerge from subsequent decision nodes" (Cole & McGinnis 2017: xvi). Action situations are either linked by *organizational* procedures, or through *rule-changing* situations: In the first case, the actors involved usually belong to different decision-units and are part of (or may even compete for) a chain of actions, where "[t]he outcomes of any one situation become inputs into the next situation" (Ostrom 2005: 56). E.g. the rule-governed competition of parties is an essential part of the democratic political process. But situations might be "only potentially linked together" (ibid. 2005: 57): Contractual arrangements or market transactions usually occur with implicit reference to an enforcement body, i.e. courts; the ease and availability of a third party to monitor, sanction and providing remedies thus impact on actors' behavior (ibid.: 57). Here, it is possible to focus on "multiple nested action arenas at any one level of analysis" (ibid.: 58); on the other hand, studying nested *rule-changing* situations demands a shift in the levels of analysis as "[a] more fundamental form of linking" (ibid.). However, according to Ostrom et al. (1994: 39), "in undertaking an institutional analysis relevant to a field setting, one needs first to understand the working rules that individuals use". As was presented above (see chapter 2.2.2), these are generated in social organization by custom or law and form the scaffolding for social organization, and hence form "the rules used by participants in ongoing action arenas" (ibid. 1994: 39).[141] However, with view to the fact that legal rules might be similar, comple-

[141] Seeing that E. Ostrom targets most of her empirical studies to the understanding of governing common-pool resources, where presumably the rules-in-use dominate social organization, Cole (2017) emphasizes the rather neglected role of formal rules within the IAD framework and how these explicitly relate to the general concept of working rules, and the rules-in-use in specific; he thus develops a typology that is presented in the following.

menting or even clash with social norms and customs, Cole (2017: 11–16) develops "a three-part typology of relations between formal legal rules and the 'working rules'" (ibid. 2017: 11):[142]

1. The formal legal rule *is* the 'working rule';
2. the formal legal rule significantly influences the 'working rule' (and sometimes vice versa);
3. the formal legal rule bears no apparent relation to the 'working rule'.

These are explained shortly in the following: (1.) In case where the formal legal rule is actually presented by the rules-in-use no conversion or interpretation is necessary, "the rule is enforced as written" (ibid.: 12); (2.) when a formal legal rule significantly influences the 'working rule', or vice versa, the degree of enforcement is rather low even though the impact of the legal rule is rather strong: Once a legal rule conflicts with a rule-in-use the former in general keeps its dominant position; (3.) in cases where the formal legal rule bears no apparent relation to the 'working rule' no enforcement is necessarily anticipated; the legal rule might e.g. constitute a symbolic act, as those "establishing an official state bird or tree" (ibid.: 13). What is outlined is a dynamic relationship between formal rules and rules-in-use, as "formal laws often influence or even determine working rules. At the same time, it is clear that social norms often influence the substance of formal laws" (ibid.: 14). In the extreme – where both formal rules and the rules-in-use are characterized by a high degree of incongruity, as shown e.g. by Theesfeld (2004: 253) – rules-in-use may create a window of opportunity for opportunistic behavior that, in turn, affect the rules-in-use and impact on the development of formal rules which might lead to its replacement (Cole 2017: 14–15). E.g. the alteration of formal rules might thus be used by the opposition for opportunistic strategies (ibid.: 16). "The implication (…) is that the various processes by which formal rules are transformed into working rules are themselves action situations, including law enforcement and other actions situations in which legal rules are evaluated and/or interpreted" (ibid.: 15). By analyzing the patterns of interactions that are the result of these rule transformations it is possible to determine (1) the very nature of the working rules in question, (2) whether these "deviate significantly from the formal rules, and (3) the extent of any such deviation" (ibid.). Consequently, "the task before us is to illustrate this interplay between institutions (rules) that are imposed on lower level entities in an economy, and the endogenous response emerging from these lower level going concern" (Bromley & Yao 2007: 8). Any alterations "in rules used to order action at one level occur

[142] The author identifies two features of Schlager & Ostrom's (1992: 250) definition of rules, firstly, a deontic element that specifies what actors may, must or must not do, and secondly, the degree of compliance/enforcement to call it a rule (in contrast to e.g. guidelines, signals or empty gestures) (ibid. 2017: 7).

within a currently "fixed" set of rules at a deeper level" (Ostrom et al. 1994: 46).[143] The analysis is hence anchored on multiple levels based on *operational*, *collective choice* and *constitutional rules* (Ostrom 2005: 58). *Operational* rules "directly affect day-to-day decisions made by the participants in any setting" (Ostrom et al. 1994: 46) and hence concern resource users, the area (where), time and frequency (when), and the way a resource is used (how), who is allowed to use the resource and to monitor its use, what information is subject to exchange among resource users, or in what ways are actions and its outcomes sanctioned and rewarded; they are located on the local level and an outcome of collective choice rules for the appropriation and provisioning of the resource, as well as the monitoring and enforcement of the rules-in-use (Ostrom 2008: 52). *Collective choice* rules affect operational choices indirectly; these rules determine "who is eligible and the specific rules to be used in changing operational rules" (ibid.); used by appropriators, officials or external authorities these are the rules in policy-making (ibid.); they hence refer, on the one hand, to formal settings of courts, regulatory agencies and legislatures on local, regional and national level, and, on the other hand, to informal venues as, e.g., gatherings and private associations (ibid.). Accordingly, policy-making with regard to the rules-in-use (to regulate so-called operational-level choices) "is carried out in one or more collective-choice arenas" (Ostrom 2005: 32; Shepsle & Weingast 1984). *Constitutional* rules are both the outcome of, and the reason for operational and collective choice rules to exist, which come into effect by (a modification of) governance and adjudication; those comprised on the constitutional level are governmental leaders, their associated parliaments and the supreme court (Ostrom 2005: 32).

Shifting levels of action is hence to be expected when e.g. decisions are made (on the operational level) about a change of rules (on the collective choice level) that impact on future actions and outcomes (ibid. 2005: 62). Thus, on the one hand, "[i]t is through shifting levels of action that participants may be able to self-consciously design rules in their efforts to change patterns of undesirable interactions and outcomes at operational or collective-choice levels (ibid.: 63). On the other hand, formal procedures, e.g. petitions or legislation, may require shifting levels of action if e.g. bureaucratic officials control the access to an arena where rules or constraints could be changed (ibid.).[144] An impediment to the participants

[143] The analyst has, for the purpose of her analysis, to decide which rules are exogenous and hence to be kept fix; once hold constant though does not mean that they cannot be changed during the course of the analysis (Ostrom 2008: 52–53).

[144] As is emphasized by Shepsle (1989), the outcome of the legislative might not be stable but may be changed (1) with a newly formed majority that supports the reversal of earlier decisions, or (2) due to changes in the constitution which may render recent changes unconstitutional. As a consequence, the author reminds that "[t]he stability of decisions in complex modern institutions is dependent not only upon the preferences and procedures used to organize decision making in one arena, but upon the entire nested set of arenas" (Ostrom 2005: 291).

to change the level of action may thus stem from high transaction costs to transform a situation which might result in sub-optimal outcomes (ibid.). Thus, the development of "de facto rules outside formal channels may be less costly than trying to use the formal channels available to participants in some political systems" (ibid.). Consequently, the assumption that multiple levels are involved in decisions that impact on lower level action arenas "greatly simplifies analysis rather than complicating it" (ibid.: 61). The study of governance and policy making usually requires the depiction of one ore more collective-choice arenas (ibid.). Depending on the makeup of the analysis and the assumptions made on the actors involved it is possible to formulate inferences on possible results (ibid.: 64). The prediction of results "is usually much more difficult (…) when one is analyzing a collective-choice or constitutional-choice level situation as it impacts on operational-level settings" (ibid.: 65).

The evaluation of outcomes is the final step in undertaking an institutional analysis. "Evaluative criteria that offer potentially practicable measures of consequence, such as some approach to maximizing a social-welfare function or some other values, whether those of Rawls, Sen, or Lasswell, are to be preferred…" (Cole 2013: 398). The task is to comparing (the distribution of) estimated costs of complete privatization with the estimated costs of alternative entitlement structures, e.g. "continued management under an amended common-property regime" (ibid. 2013: 383).

4 AGRICULTURAL LAND PRIVATIZATION IN GEORGIA

To reach the sky one needs a ladder.
To cross the sea one needs a bridge. To reach mankind needs justice.
(A hazelnut farmer in Kulishkari, Samegrelo, July 9, 2013)

4.1 From the plan to the market: Georgia's transition

Georgia is inhabited by approximately 4.7 million people, with about half of the country's populace living in rural areas (Kan et al. 2006: 2). As an area of peaks and valleys, with 69,500 km^2 of mountains covering 54% of the total surface;[145] about 43% of the total area is taken by agricultural land (3.02 million ha) and approximately the same area is covered by woods (Ebanoidze 2003: 126). For agricultural purposes, pasture land has the largest share at 25% (1.8 million ha), arable land constitutes 11.5% (0.8 million ha), perennial crops 4.3 % (0.33 million ha) and meadows comprise 2% (0.14 million ha) of the country's overall territory (ibid.; see figure 4-1). Kan et al. (2006) stress that "a large part of the agricultural land in Georgia is in green mountain pastures, and only 40% is suitable for cultivation" (Kan et al. 2006: 6).

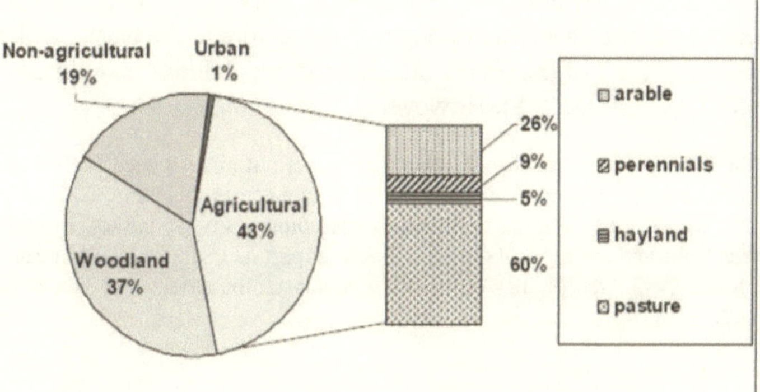

Figure 4-1: Land structure in Georgia (Salukvadze 2006: 6)

Georgia's agriculture has a rich history and represents a primary economic sector due to its diverse climate and soils (Didebulidze & Urushadze 2009: 241). In Soviet times, agriculture generated high-value export products, such as tea and citrus as well as "wine, other alcoholic beverages, fresh and processed fruits and vegetables, essential oils and spices" (Didebulidze & Plachter 2002: 87).

[145] 69,500 km^2 = 6,950,000 ha.

There are "over 500 varieties of native grapes (…) still grown in Georgia, which produced most of the quality wines in the Former Soviet Union…" (Bezemer & Davis 2003: 8). Besides of being a supplier for food products and minerals, Georgia was the "center of tourism for the centralized state economy" (Slider 1995). Consequently, the Georgian populace enjoyed the highest living standard within the Soviet Union (GTAI 2015: 4). Production was organized in large-scale state-owned (*sovkhoz*) as well as collective (*kolkhoz*) farms that were centrally managed and controlled (see below);[146] rurally located families additionally cultivated a plot of 0.25 ha for their own production, the so-called *household plot* (Ebanoidze 2003). With the dissolution of the Soviet Union, agricultural output decreased dramatically, as in other republics of the Soviet production and distribution system (WTO 2009a: 62). Due to this collapse of the former system,

> …a largely unprepared and ill-equipped farming sector had to assume the role of the country's chief guarantor of food security. Most farm families lacked basic knowledge of running an independent farm enterprise, and were unfamiliar with some of the techniques of cultivating basic food crops. As a result, food production by 1994 was 60% of its mid-1980s levels. (Bezemer & Davis 2003: 13)

By 2013, "…about 54% of the active workforce was employed in agriculture in Georgia, 80% of which are self-employed, meaning that most Georgian farmers were engaged in subsistence farming, producing products mainly for their own consumption and using outdated tools and methods" (Bluashvili & Sukhanskaya 2015). Moreover, agricultural production in the country depends "on irrigation infrastructure in the east and drainage infrastructure in the west[, which] virtually collapsed as a result of the civil war, vandalism and deferred maintenance" (ibid.: (Bezemer/Davis 2003: 7–8). However,

> [t]he relatively low level of cereals output has not always been sufficient to satisfy domestic demand. This is a consequence of Soviet central planning, which decreed that farms in the west of Georgia, traditionally the country's bread basket, should specialise in citrus fruits and tea and that Georgia should import its grain from Ukraine and Kazakhstan. Since 1992, Georgia has been receiving substantial amounts of wheat as food aid. (ibid. 2003: 8)

[146] The differences between a *sovkhoz* and *kolkhoz* lie in their political function and their subsequent property division: Whereas both the means of production and output of a sovkhoz was state-owned, the kolkhoz' means of production was under state ownership (land) as well as owned collectively (machinery, installments, etc.), in what was known as an *artel*, while the output was commonly shared; remuneration in the former case was paid by the state in the form of wages, whereas kolkhozniks received payments in cash and in kind (Vucinich 1952: 102). Even though their employees performed the same work, the former counted as workers and were closely connected to the Party and the state (and, hence, included in the social security system), whereas the latter were perceived as peasants and thereby "not [yet] full-fledged members of the socialist community" (ibid. 1952: 102).

The Soviet land administration and agricultural production system

In the initial phase of Russian and, later, Soviet influence, several methods were applied to gain state control over privately owned land, namely "expropriation, nationalisation and collectivisation" (Ho & Spoor 2006: 630). During the Soviet era, a hierarchical land administration system was established using a top-down approach, where the locus of state control "was forcibly switched from a concern of legal ownership of land to a concern for land use and land capability" (ibid.: 630). With the aim of easing centralized, large-scale management and resource allocation within the Soviet Union, multiple agencies were set-up, and

> each charged with strict responsibilities for administering particular land resources (soils, water, forests, minerals etc.). While these agencies produced and kept excellent records of land use and resources (such as residential buildings by the Bureau of Technical Inventory…), this information was spread amongst a number of increasingly independent and isolated agencies at the time of break-up. (ibid.)

Especially in rural areas, this resulted in physical changes in landscape and land use (ibid.). For example, "[c]ollectivisation of agricultural land […] led to a forced consolidation of land parcels into large-scale field blocks" (ibid.), where traditional physical boundaries were removed and replaced by collectivized agricultural land that was necessary for the kolkhoz and sovkhoz production system.

The characteristics of the Soviet agricultural production system have been summarized as follows by Bromley (2008a: 228): (1) low agricultural productivity; (2) not economic but rather political allocation of agricultural inputs and machinery, which hampered timely provisioning; (3) incentive problems within agrarian production units that led to the reallocation of inputs and labor for people's own needs; (4) loss of produce due to a defective agricultural marketing system; (5) as a result of Stalin's resettlement policies, which first destroyed and then newly created towns and villages in the form of production sites, the latter became the sole provider of social services such as housing, schooling or transport and, hence, "represented a curious mix of industrial-scale farming and required civic functions. Those two purposes often competed for scarce labor, financial resources, and managerial talent" (ibid.: 228).

Land reforms initiated after Georgia's independence

Land reforms were introduced after Georgia's independence in 1992 in the aftermath of a coup d'état, though inter-ethnic wars in the breakaway regions South Ossetia and Abkhazia as well as civil unrest in Samegrelo foiled establishment of a steady domestic political environment before 1998 (cf. Cornell et al. 2005; Wheatley 2003; Halbach & Müller 2001). Poverty in post-Soviet Georgia is centered in rural and urban areas outside of the capital (Reisner/Kvatchadze 2005: 19). Although real incomes increased between 2002 and 2009 by 6% in urban areas, they exhibited a decline of 3% in rural areas (UNDP 2009).

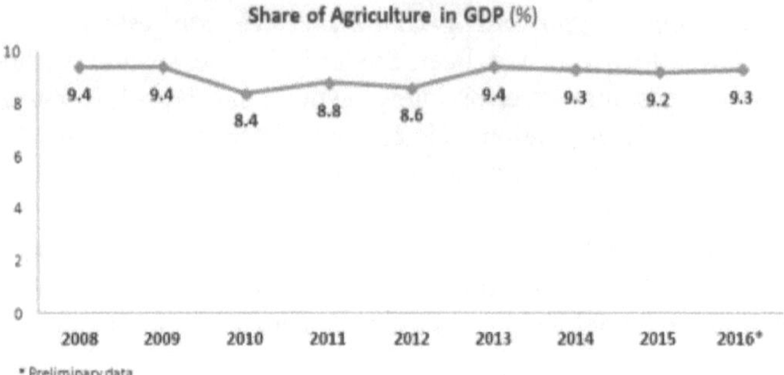

Figure 4-2: Share of Georgian Agriculture in GDP (GeoStat 2017)

While agriculture accounted for over 30% of GDP in 1990, it was only less than 10% of GDP between 2008 and 2016 (ibid. 2009; GeoStat 2017, see figure 4-2).

Since the mid-1990s, the Georgian Government has received heavy assistance from the European Union, World Bank and IMF (Papava 2003). From the Rose Revolution in 2004 to the August War in 2008, the country's GDP increased by about 9% per year (World Bank 2009). At the same time, however, poverty and extreme poverty did not show any change at all (Welton et al. 2008: 34). In fact, it was estimated in 2010 that 55% of the Georgian populace lived on less than $2 per day (gtai 2010: 2).

The emergence of institutional arrangements vis-à-vis land in the newly constituted transition economies was either dominated by state-led or market-based reforms. In the former case, the disaggregation of land took place within historical boundaries by restituting land to former owners (e.g. Czech Republic, Baltic states) or by allocating land shares on a voucher basis (e.g. China, Uzbekistan), where land remained under state ownership and usufruct rights to land were allocated. For cases of market-based reforms, land was both sold and leased-out (e.g. Poland), or ownership rights to a limited number of physical plots were distributed to rural households (e.g. Albania and the South Caucasian republics; (Hagedorn 2004: 4–5; Swinnen & Heinegg 2002).[147] Restitution was avoided in some countries "because there was no possible return to the feudal structure that preceded collectivization, land had been fully expropriated in the transition to socialism..." (Janvry & Sadoulet 2001: 16), whereas geographical changes, ethnic or equality concerns as well as (the promotion of) certain norms and values were further reasons why former ownership rights were not or could not be enforced in others (Hagedorn 2004).

In Georgia, the society had been subject to a system of feudal tenure and was deeply hierarchically organized when it became part of the Russian Empire in

[147] Janvry and Sadoulet outline how individualization, i.e. the distribution of individual family farms, was the favored option to allocate land among countries "where the productivity of the collective farms was the lowest and where the share of agriculture in total employment was the highest" (Janvry & Sadoulet 2001: 16).

1801, until the Bolsheviks' takeover in 1917 and the ensuing abolition of private ownership (Suny 1994). Invasions by Mongol, Persian and Ottoman troops in the period from the 13th century to Russia's annexation in 1801 – the so-called "long twilight" – had led to its division into kingdoms (Imereti, Kartli, Kakheti) and principalities (Abkhazia, Guria, Mingrelia, Svaneti, Samtskhe), which existed as separate units (ibid. 1994: 74-76; Waters 2004). Almost two centuries later, former ownership rights in land were hardly enforceable for the majority of areas.[148] After independence in 1991, agricultural land plots were distributed foremost to rural households in addition to the common household plot of 0.25 ha, as " …the Georgian move to distribute land focused only on one component of the sector – the subsidiary household plots – and did not propose any program for the traditional farms" (Lerman 1996: 1). Market-based reforms to privatize the remaining land under state ownership were introduced in Georgia in the mid-1990s, guided by Washington-based international financial organizations, such as the World Bank or International Monetary Fund (IMF), which advocated "development strategies focusing around privatization, liberalization, and macro-stability (meaning mostly price stability)" (Stiglitz 2008: 41), articulated via the policy prescriptions of the Washington Consensus (Williamson 2005, 1990; Stiglitz 1999).[149] According to Stiglitz (2008), it was "a set of policies predicated upon a strong faith – stronger than warranted either by economic theory or by historical experience – in unfettered markets and aimed at reducing, or even minimizing, the role of government" (ibid. 2008: 41; Heynen et al. 2007).[150] Subsequent policies relied on the idea of trickle-down effects – meaning that confidence that a

[148] According to L. Tchanturia, Head of Administration at GTZ Regional Office South Caucasus, Tbilisi (from 2006–2007 former Head of Registration and Cadastre Unit of the NAPR) (personal communication, July 24, 2014).

[149] The term was initially coined by Williamson to refer to "a list of ten policy reforms that were (…) widely held in Washington to be needed in most or all Latin American countries as of 1989" (Williamson 2004: 3). He stresses that the term "consensus" is oversimplified, for even within these organizations critique was widespread (ibid. 2004: 2; cf. Williamson (2008).

[150] First-generation reforms targeted instruments and inputs needed to realize "objectives", such as privatization, tariff and budget cuts or macroeconomic stabilization; in contrast, the agenda was "augmented" in the mid-1990s by second-generation reforms that were rather outputs, "statements of desired outcomes (e.g. civil service reform or improving tax collection), without a clear sense of policy design" (Navia & Velasco 2003: 266). The latter have, moreover, been characterized as "a motley crew, encompassing broad reforms of the state, the civil service, and the delivery of public services; of the institutions that create and maintain human capital (e.g. schools and the health care system); and of the environment in which private firms operate (more competition, better regulation, stronger property rights)" (ibid. 2003: 266; cf. Rodrik (2004, 2002)). Navia & Velasco (2003: 266) point out that whereas "the 'victims' of the first-stage reforms were often atomistic or too poor to matter politically (…), the set of interests potentially affected in the next stage reads like a Who's Who of highly organized and vocal groups: teachers' and judicial unions, the upper

rise in average incomes would lead to increase of those of the poorest – and were implemented as "clean model institutions" (Stiglitz 2008: 47), while distributional aspects were left aside in the belief "that somehow everybody would benefit in the way that a rising tide would lift all boats" (ibid. 2008: 47; cf. (Krugman 2008). As evidence suggests, this was not necessarily the case, and the Russian financial crisis that emerged in 1998 serves as prominent example (Stiglitz 2002).[151] A report by the Working Party on Land Administration of the United Nation states accordingly

> ...that the rate of progress was slower than required to facilitate development of market-oriented economies, and that the high failure rate of many projects was the result of underestimation of the scale and complexity of reintroducing private ownership (United Nations Economic Commission for Europe (UNECE, 1999). (Ho & Spoor 2006: 630)

As is also highlighted by Bromley (1997), the privatization of land in the formerly planned economies was a "process of creating markets where none existed before" (Bromley 1997: 7). The aim was "to create private property rights in natural resources and then to 'let the market work" (ibid. 1997a: 7). The outcomes of this liberal approach in Georgia are outlined in the following sub-sections.[152]

4.2 State-led reforms (1992–1996)

The distribution of agricultural land officially commenced in January 1992 with the passing of Resolution Nr. 48 of the Cabinet of Ministers, according "to which agricultural land within the borders of Georgia was to be transferred to the ownership of the citizens of Georgia" (Gvaramia 2013: 5) For that purpose, "local commissions were established to redistribute the available lands in the *rayons* [districts]. The process was only completed in 1998, due to the turmoil in the

---

echelons of the public bureaucracy, state and local governments, owners and managers of private monopolies, and the medical establishment".

[151] As outlined by Williamson (2000a: 610), "[i]n the confidence that the future would take care of itself, the mass privatization program that was begun in the spring of 1992 had purportedly reached a "triumphant completion" in June 1994 [...], by which date two-thirds of Russian industry was privately owned". Though "[g]reater appreciation for the shortfalls of the institutional environment in Russia would have led to more cautious pronouncements (Anders Aslund 1995)" (ibid. 2000: 610). Critique of the agenda of the Washington Consensus is provided by Stiglitz (2008; 1999), Williamson (2008, 2005) and Gore (2000).

[152] The idea of the supremacy of "the market" over governmental "interference" by laws and regulations is supposed to either stem from a purely economic belief understood as rational response to formally excessive state interventions or from "an ideological position that government is necessarily bad and that volitional bargains are necessarily good" (Bromley 1991: 35). A critique of the market as a social artifact is emphasized by e.g. Bromley (1989: 111): "It is the structure of conventions and entitlements that forms the legal foundations of the market and its commodity transactions", and hence "requires purposeful actions of the state to permit the establishment of market processes" (ibid. 1989: 63; cf. Bromley (1997b).

Georgian regions" (Lohm 2006: 29), because of which reforms for the distribution of land

> …to rural households were intended to satisfy local subsistence needs and to ensure a fairly regular flow of surplus food commodities to urban markets in a time of general unrest and civil strife. The land distribution strategy has been often credited with enabling Georgia to avert widespread famine during the early years of disruption and civil war. (Lerman 2005: 1)

The forming of a subsistence sector was to be realized by allocating 0.8 million ha of arable and perennial "reserve land" in a privatization fund by locally elected land reform committees who distributed plots between 0.15 and 0.5 ha of agricultural land at no cost to rural households according on the following criteria (Ebanoidze 2003: 126):[153] In the lowlands, former kolkhoz or sovkhoz staff received 1.25 ha per household, other rural households got 0.75 ha with the exception of some eligible families living in the highlands, who received up to 5 ha; meanwhile, urban settlers obtained 0.25 ha agrarian land (ibid. 2003: 126). According to the State Department for Land Management (SDLM), land allocated through the "privatization fund" add up to 30% of the country's overall agrarian area and 60% of the total arable and perennial land (ibid.: 126). In 1993, the Cabinet of Ministers issued Decree No. 503, authorizing local land reform committees to hand out so-called Receive–Delivery Acts (also known as Shevardnadze documents), which "are considered as the main document for granting ownership of agricultural land to households. But these certificates confirmed the right to use, not ownership" (USAID 2011a: 7). Moreover, due to their high cost, "the majority of new owners did not obtain Receive-Delivery acts… In addition, the government was unable to finance the survey activities and preparation of other legal documents necessary for registration of ownership to the land" (Ebanoidze 2003: 126). Consequently, "[t]he high price of the documentation which was calculated to cover the costs of the privatization process was an important reason why 70% of the population had not received the Acts by 1998" (USAID 2005: 7).

By 1996, almost four million parcels – constituting 0.93 million ha, which was approximately 15% more land per household than foreseen – had been distributed to 1.4 million households. As "allocation took place based on outdated and often incomplete land survey records, the area of the allocated land parcels often did not correspond precisely to the norms set out by the Resolution of 1992" (ibid.). According to "the framework of the reform, agricultural land would be provided to households (…) although there were also cases where land was granted to individuals which occurred in obvious violation of the terms of the reform" (Gvar-

[153] "A household is defined as a family living in a village, community or other similar settlement or a family living in a rural area and engaged in agriculture" Gvaramia (2013: 5). During Soviet times, "almost all rural families cultivated a household plot which was used to augment the family diet" Wegren (2002: 1034).

amia 2013: 5). Moreover, "a number of cases of the oppression of ethnic minorities took place during the first stage of Georgia's land reform" (ibid. 2013: 7), "with limitations placed mainly upon the Kvemo Kartli and Samtskhe-Javakheti regions owing, most likely, to the large non-ethnic Georgian local populations" (ibid.: 6; cf GTZ/CIPDD 2006), namely ethnic Azeria and Armenian minorities.

The transfer of ownership to actual users was approved by the Law of Georgia On Agricultural Land Ownership (1996), "more than four years after distribution of land had begun" (Lerman 1996: 1). The Law stipulated "that land ownership was a privilege of citizens and households of Georgia as well as legal entities registered in Georgia and operating in the agricultural sector" (Gvaramia 2013: 5).[154] The Law also included "[h]ousehold land, kitchen gardens and land attached to summer cottages of certain defined areas" (ibid. 2013: 5). Another Law on Land (Immovable Property) Registration, enacted in the same year, stipulated that "in order to become privately owned, land has to be registered and registry certificates have to be issued" (ibid.: 12).[155] But the existing paper-based archive system "only recorded initial owners" (Ebanoidze 2003: 128) and, thus, did not track further land transactions. Consequently, "[i]n order to keep their power and influence, the bureaucrats slowed down the process" (Bezemer/Davis 2003: 12). For example,

> ...the Ministry of Agriculture and Food Industry (MAFI) attempted to block and subvert the US Agency for International Development (USAID) Land Registration programme, as this programme would speed up the transfer of control over land from bureaucrats to beneficiaries. (ibid. 2003: 12)

Financial obstacles were created to slow down the process at a time where an average monthly salary in Georgia hardly exceeded GEL 47 [USD ~ 19] (ibid.: 12).[156] In particular, "USAID insisted that Registration Certificates should be issued for a nominal one GEL [0.40 USD] fee. The Ministry demanded a 1000 GEL [405 USD] fee" (ibid.). At the same time, "[p]opular pressure for the transfer of property rights and registration of land plots was not high" (ibid.: 13). Rather, "[t]he beneficiaries had insufficient information to allow them to estimate how restricted their rights in land were or to appreciate the need for registration" (ibid.). For the majority of the newly created landowners, land allocation still appeared to lack any legal guarantee (Ebanoidze 2003: 126). It took a further three years until a new law was passed intended to regulate the registration of immovable property as, in April 1999, the Georgian Parliament passed the Law On Land Parcels State Registration Fees with the aim "to complete the process of land reform by 2003" (Bezemer/Davis 2003: 13).

---

[154] "Privilege" here should not be confused with Hohfeld's (1913; 1917) legal correlates; the law targets ethnic Georgian citizens and households which was deemed unconstitutional in 2012 by the Constitutional Court of Georgia (see chapter 4.3).

[155] It should be emphasized that no further information could be obtained on the Law on Land (Immovable Property) Registration (1996) than from Ebanoidze's (2003) report.

[156] 1 GEL = 0.404984 USD (September 20, 2017).

4.3 Market-based reforms (1996–2012)

In 1996, the Law on Agricultural Land Leasing initiated a market-oriented land reform to lease land already under state ownership to natural or legal persons for a maximum period of 49 years (Tsomaia 2003a: 9).[157] Responsible were municipalities acting through local SDLM offices; on the national level, state-owned agricultural land was directly leased for a small number of cases (USAID 2011: 7). This process was, however, reportedly biased and corrupt: "Leased land tended to be the larger, and some considered better quality, arable land parcels which were retained by the state after small scale privatization" (ibid. 2011: 7). Particularly of note are reports that "[i]nfluential government officials received the most fertile parcels, even though they had neither experience nor interest in farming" (USAID 2010: 10).[158] The process of leasing-out land has been characterized as having

> ...many flaws, including favoritism and lack of transparency in the allocation process and lack of protection against changes of local government. Leases were not respected as legal rights *in rem*, but were considered to be mere contractual rights subject to revocation by the locality. (USAID 2011: 7; italics in the original)

It has been estimated that by 2002 the major share of leased land was actually being held for economic (sub-renting) or speculative purposes (ibid. 2010: 10).[159] At that time, the allocation of leaseholds was distributed as follows: More than 90% of leases were held by natural persons, with only about 5% of individuals controlling 50 ha or more, amounting to 24% of all arable, namely cropped land, whereas 62% of all leases by individuals were for less than 10% and approximately 30% of individual leases held 3 ha of less of cropped land and, thus, only controlled 2% of total cropped land, In contrast, legal persons, such as limited liability companies, stock companies and the like, were relatively few in number but controlled the majority of leased cropped land: 8% of such lessees leased more than 50 ha per leasehold and controlled roughly 56% of total cropped land (Tsomaia 2003: 8).

As a result of this process of leasing, by 2002, it is reported that about 25% of agricultural land in Georgia was under private ownership. This newly privatized agricultural land mostly "lies in the vicinity of settlements and in Georgia's more important agro-ecological zones. The majority of the land designated as arable (55% of total arable) or used for perennials (68% of total perennials) was privatized" (Egiashvili 2002: 2). As a consequence, "[t]he remaining state owned land is mainly to be found in remote, often mountainous areas where there is a clear

---

[157] The Law on Agricultural Land Leasing (1996) was repealed by the Civil Code in 1997 and brought back into effect by Presidential Decree (No. 446) in 1998 (Tsomaia 2003a: 9).

[158] Tsomaia (2003b) provides the latest numbers on land under leasehold as well as under state ownership (USAID 2013).

[159] It is assumed that 25% of total arable land (cropped areas) was privatized, while 30% (10% of total agricultural land) was leased out (Tsomaia 2003a: 4).

dominance of pastures" (ibid. 2002: 2), but "often with very limited accessibility" (ibid.: 2). Moreover, Egiashvili (ibid.: 2) stresses that although

> … 75% of land [being] in state ownership might at first glance indicate that land is available, the actual situation is that the "valuable" land in the vicinity of villages is either privatized or leased, so that not much reserve land is available in the agriculturally important areas.

Further reforms to privatize state-owned agricultural land were initiated after the Rose Revolution in 2005, with the Law on State-Owned Agricultural Land Privatization regulating privatization of formerly leased land as well as unused or vacant land intended to complete the privatization process in the following manner:[160] First, leaseholders were granted the right of preemption and "given a 5 year window in which to directly privatize their holdings which ended in May 2011" (Tsomaia 2003: 8). Second, a new payment scheme allowed for "10 year terms for leased land privatization payments, accompanied by discounts for those who chose to pay by lump sum" (ibid. 2003: 11).[161]

The public auctioning of unused (vacant) state-owned agricultural land was initiated during this time. Known as *privatization through special auction*, it was conducted in the following way: For those lands that leaseholders were not interested in buying, an auction was held, open only to Georgian natural persons and legal entities (registered in an *estate book* kept by the local municipality, *sakrebulo*). Lands not sold at these auctions could be acquired in a second auction that was open to any bidders (USAID 2011a: 10).

The Law on State-Owned Agricultural Land Privatization, Art. 7 further specifies that, in

---

[160] Moreover, the LMDP supported "the operation of one central and 5 regional GIS support centers which checked boundaries of the leased parcels prepared by surveyors hired by lessees against the integrated cadastral and aerial images prepared by LMDP to ensure there was no overlap with other parcels in the area. An additional 5 GIS support centers were established on a temporary basis to meet demand. The centers provided the services free-of-charge to leaseholders, surveyors and government officials" (USAID 2011b: 9).

[161] The Tax Code (2004) charged local administrative units with rules to calculate the land tax and lease rates, ranging from 8–51 GEL per hectare of cropland and GEL 3 per hectare of pasture lands annually (Tsomaia 2003: 6). However, Tsmomaia et al. (ibid.) underline that calculation and paying of lease rents, which were calculated to be at least equal to land taxes, have been rather vague in practice. In 2011, the Government of Georgia set (and increased) the annual rates of property taxes on agricultural lands (APLR 2011). In addition, local authorities were given the right to raise the tax by 50% (ibid. 2011). Tax exemptions apply to land plots not exceeding 5 ha, or to "[n]atural persons or legal entities who have received agricultural lands for re-cultivation purposes - for the first five years following its allocation" (ibid.). In 2010, a "new Tax Code abolishe[d] taxes on transactions in property, 0% profit tax and 0% VAT; 0% VAT on primary supply of agricultural products; and 0% import duty on agricultural and other equipment" (WTO 2009b: 64).

> …privatizing the state-owned agricultural land by a special auction, open auction, or the leased land through direct sale, the territorial agencies of the Ministry of Economic Development of Georgia shall work out a land and other real estate purchase deed, which represents the grounds for entering the property right in the Public Register.

According to a USAID (2011a) report, the acquisition of formerly leased lands was characterized by "fraud with several limitations and complications. Former lessees reported that they had to pay fines and taxes on lands they were attempting to purchase back and that it seemed ad hoc and disorganized" (ibid. 2011a: 3-4).

In 2007, the 100 Agricultural Enterprises initiative was announced by the President of Georgia, with the aim of privatizing more than 40,000 ha of land consisting of plots of more than 50 ha to both domestic and foreign investors *under specific conditions*, such as the buyers setting up processing plants (USAID 2011b: 16; Civil Georgia 2007a):[164] This turned out to be quite a bargain for those who could meet such conditions, as "processing companies benefit[ted] from preferential terms and [could] acquire these estates for only 20% of their market price" (Cordonnier 2010: 5). A contracting collaboration between APLR, on the one hand, and the MoE and MoA, on the other, led to the surveying and preparation of documentation "on an additional 5,192 hectares of land in the Kvemo Kartli, Khakheti, Samegrelo regions" (USAID 2011b: 16) from the end of 2008 to February 2009. Even though the project objective of creating one hundred enterprises was not met, "the program resulted in sale of thirteen large land parcels comprising 7,318 hectares, the largest being 2,323 hectares and the smallest being 10" (ibid. 2011b: 16).

Further privatization measures to sell off state-owned agricultural land have been in the form of *direct sales* (which could include a free-of-charge transfer), based upon favorable decisions of the President of Georgia resulting from a competitive selection process as well as direct sales of leased land, even though the latter "has not been in force since 1 May 2011" (Gvaramia 2013: 10).[165] However,

---

[164] In his annual State of the Nation Address on February 11, 2011, the president declared that the "overall situation in the sector remains very dissatisfactory, including in creation of new agricultural processing enterprises" (DWVG 2011b).

[165] Generally, an *auction* "is a process of trade during which participants are ready to give their price to win the competition. The winner is the person who proposes the highest price" (Gvaramia 2013: 11). Accordingly, "[t]he winner of the auction with special conditions is a person who takes the responsibility for satisfying the conditions and offers the highest price to the agency implementing the privatisation" (ibid. 2013: 11). *Direct sales* are based on "the decision […] by the President of Georgia based upon the proposal of the Ministry of Economy and Sustainable Development and, in special cases, the Government of Georgia" (ibid.: 11); decisions on *privatizations that include free-of-charge transfers* are "made by the President of Georgia. This is for the citizens of Georgia who lived or still live on occupied territories and remain homeless. The proposal about such privatisation is prepared by the Ministry of Economy and Sustainable Development and is submitted to the President" (ibid.); finally, *direct sales in a competitive selection process* "is one of the forms of privatisation when a decision is made by the President of Georgia and property

the "most popular form for selling agricultural land" in the period 2011–2013 was actually "[e]*lectronic auctions*" (ibid. 2013: 11).[166] The initiative for an auction to acquire state-owned agricultural land "can be initiated by a citizen of Georgia, a legal entity of private law registered in Georgia as well as third parties" (ibid.: 14). In the next step, decisions to hold auctions to privatize state-owned agricultural land are

> made by the appropriate agency in charge of implementing privatisation; namely, the Ministry of Economy and Sustainable Development and its territorial units or agencies to which the Ministry has delegated such authority. The responsible agency reviews applications within a two month (although there are cases when such a process lasts for three-to-five months) and makes a decision on the privatisation of the agricultural land plot through auction (announced mainly on-line). (ibid.: 14)

In 2007, the Georgian Government targeted the "recognition of property rights to lawfully possessed (used), as well as squatted land" (Law of Georgia on recognition of property rights of the parcels of land possessed [used] by natural persons and legal entities under private law, Art. 3). In cases of unauthorized occupation or lack of documentation proving legal ownership, "the power to grant legal ownership is held by the Property Rights Declaration Commission established at the local self-governing administrative unit" (TI Georgia 2012b).

In 2005, pasture land was defined as municipal property under the Law on Local Self Government (2005), and municipalities (sakrebulos) were instructed regarding the contracting of lease agreements. In 2010, however, pasture lands were retransferred to state ownership, since "local municipalities failed to successfully manage pastures" (USAID 2010: 6). Pastures that were leased prior to July 30, 2005 were now subject to privatization, and lessees had to purchase them by May 2011 at latest. The remaining pastures were leased directly or through auctions by the Ministry of Economy and Sustainable Development (MoE), which is also in charge of privatizing lands (ibid.: 7):[167] "APLR facilitated in privatizing of pastures under leasehold. In Marneuli the most of pastures, ~ 80%, is sold out to

---

rights are granted to interested parties (potential investors) that will fully and duly comply with conditions set for a competitive direct sale of agricultural land. *Direct competitive sales* of agricultural land are implemented in the case of the existence of numerous conditions for investment and when the terms offered by the interested parties significantly differ" (ibid.).

[166] An electronic auction is "held with or without special conditions and will identify a winner as the person offering the highest conditions. The electronic auction is held through a webpage (…). Identities of participants are confidential" (ibid.: 11). The electronic auction is currently available at https://www.eauction.ge/. For further details on announcing and participation in an auction, see Gvaramia (2013: 14). The so-called "start-up privatisation price is approved according to Resolution No. 15 of the Government of Georgia of 13 January 2011" (ibid. 2013: 14).

[167] As many of the sources used for this thesis appear to have been written by non-native English speakers, there are a number of deviations from common English usage in the quotations

private entities/persons. Only large pasture lands are not privatized; they mostly are in communal tenure" (USAID 2011b: 82). The final outcome of these efforts is, however, not known today. As can be read in another USAID (2013: 7) report,

> …it is still hard to find out what size of land belongs to the private sector in Georgia, what is the total size and the number of such land parcels. (…) More importantly, even state agencies find it hard to obtain the information about what size of agriculture land is under the state ownership.

Until 2010, the agricultural sector was to a large extent funded by international donors (DWVG 2011d), and reforms took place in an "ultra-liberal economic environment" (Civil Georgia 2010a). As is assumed by Cordonnier (2010: 4–5),

> …the main reason of the bad performance of Georgian agriculture results from a combination of post-transition fragilities (small scale of plots, destruction of previously existing value chains), and of the rather "naïve" approach of market liberalization followed in agriculture by the Georgian government since the Rose Revolution (…). By implementing a large scale reduction of tariffs after the Rose Revolution, including in the usually highly protected food sector, Georgia has forced its farmers to compete with both highly subsidized and/or highly efficient farmers of neighbor EU and of other leading agricultural countries (Ukraine, Russia, and Turkey).

Though Georgia had been called the "beacon of democracy" by President George Bush, Jr. in 2005 (Halbach 2007: 1), President Saakashvili's reign (2004–2012) increasingly took flak for authoritarian and libertarian rule (Jobelius 2012; Narmania 2009; Halbach 2008).[168] In 2010, the government began promoting land privatization extensively by trying to attract land acquisition from foreign investors. For example, at the end of 2010, the first such program was set in motion by inviting South African Boer farmers to purchase land in Georgia, which was accepted by a few individuals who acquired land (Prasad 2012; DWVG 2011c). Within the ensuing two years, a number of cases were reported of the eviction of Georgian farmers from their privately-owned arable lands (Georgien Nachrichten 2010, 2011; Corso 2012) as well as the revocation of communal and leased pasture land (Gegeshidze 2012: 35; TI Georgia 2014a). A prominent case of land purchase by a foreign investor is Ferrero, which "owns hazelnut plantations all over the world and is the second-largest foreign agricultural land owner in Georgia (…) it currently owns and farms approximately 3,500 hectares" (ibid. 2014).[169] Nevertheless,

---

used throughout. Rather than unduly altering them or indicating my awareness of these language issues with the conventional *sic*, I will leave them as is, unless they seem likely to lead to misunderstanding.

[168] An amendment to the Georgian Constitution after the president's inauguration in February 2004 weakened the role of the parliament in the political process, a move toward "super-presidentialism" (Freedom House 2005).

[169] Ferrero, the second largest land-owner in Georgia (TI Georgia 2014a), is part of the case study on hazelnut in Samegrelo (see chapter 6.2.2.1).

> …the company has been hindered by the lack of a comprehensive national land registry: 30% of the 1,200 hectares of land that Agri Georgia [a Georgian subsidiary of Ferrero] purchased from the Ministry of Economy between 2009 and 2011 were already occupied and being farmed by local farmers, and some of this land included local cemeteries. Agri Georgia gave these occupied land titles back to the state and purchased an additional 1,200 hectares. The company had to create its own internal land registry, demarcating land boundaries with GPS – a process that took two years. (ibid.)

Increasing arrival of Punjabi farmers in Kakheti, East Georgia, resulted in an official petition to revoke the practice of selling land to non-locals (IberiaPress 2013b).[170] In 2012, the Constitutional Court of Georgia settled the case of a Danish citizen against the Georgian parliament, disputing the right of non-Georgians to land ownership in favor of the citizen of Denmark (Patsuria 2012). Moreover, "the Plenary of the Constitutional Court of Georgia ruled that several regulations of Article 4 of the Law of Georgia on Ownership of Agricultural Lands (2005) were unconstitutional and, therefore, void" (Gvaramia 2013: 11). Thus,

> … a foreign citizen was only allowed to become the owner of agricultural land in Georgia if such land was inherited or had been previously rightfully owned as a citizen of Georgia. At the same time, foreigners and legal entities registered in foreign countries were to transfer agricultural land under their ownership to Georgian citizens, households or legal entities registered in Georgia within six months after such ownership was initiated. Nevertheless, the Parliament of Georgia has not adopted any changes to the Law of Georgia on Ownership of Agricultural Lands at this stage". (ibid. 2013: 10)

In general, "[o]rganized information about land use is (…) scarce, which is a major obstacle to designing and implementing a sound agricultural and land policy" (Egiashvili 2013: 14). Concerning the privatization of agricultural land in Georgia, the following challenges remain: Bureaucratic processes are slowed down by the respective agencies in charge of privatization, as a number of cases have revealed that "initiators of requests for the privatisation of agricultural land wait for the announcement of the auction for several months before it takes place…" (Gvaramia 2013: 18). On the other hand, a "*serious challenge is a lack of awareness on the part of the population on the procedures for privatisation*" (ibid. 2013:

---

[170] Transparency International Georgia estimates that about 0.7 per cent of all agricultural land belongs to non-locals (TI Georgia 2014a). Regarding Indian farmers, there was an estimated "200 to 250 Indian farmers working in Georgia [in 2014], down from a previous record of 2,000 immigrants" (ibid. 2014a), who had already moved on to Armenia and Uzbekistan; in the latter case, "their assistance in helping local farmers has been specifically requested by the Uzbek government" (ibid.). One case is reported of an Indian farmer who "purchased three hectares of agricultural land in Sagarejo in February 2013. (…) In total, he estimates that he has invested GEL 10,000, in addition to purchasing the land. In January 2014, [he] (…) applied to the State Services Development Agency for a renewal of his visa but the agency refused, saying that he did not receive a recommendation from the Ministry of Internal Affairs. (…) [He] is preparing to file a claim against the Georgian government to seek compensation for his financial loss" (ibid.).

18; italics in the original). Consequently, "[t]here are frequent cases when the population does not or cannot participate in the auctions simply because of their lack of knowledge of the procedures for which they cite different reasons" (ibid.: 18). Another challenge is the development of broker services "with only a small number of individuals in the country dealing with these issues and without the existence of qualified brokers (companies) working with agricultural land" (ibid.).

In 2013, the parliament set up a sectoral program for those farmers having registered their land to support land cultivation by the provisioning of vouchers to farmers owning land up to 1.25 ha (DWVG 2013: 2).[171] My analysis in chapter 5 seeks to shed light on the realization of this process of distribution of vouchers to land owners who had registered their property (see chapter 6.1).[172]

4.4 Donor-led land tenure formalization (1996–2012)

The theoretical foundation "and justification for land titling, land registration and land administration projects that formalize property rights are based on the privatization approach" (Loehr 2012: 837), which gained support among the international donor community beginning in the 1990s (see chapter 4.1). As a consequence, "[t]he number and size of projects designed to formalize property rights in developing, threshold and transitional countries have increased exponentially during the past decades, e.g. in Thailand, Indonesia, Philippines, Ghana, or Bolivia" (ibid. 2012: 837). The *formalization of land tenure*, i.e. the institutional transfer from paper-based to digital, geo-referenced property transactions, which involves titling, registration and the administration of land, has been one of the key approaches for securing individual land rights, stimulating the development of land markets, and encouraging production (ibid.).[173]

---

[171] Besides, plans are elaborated for the support of cooperative farming (DWVG 2013b: 2). The World Bank is intended to start an irrigation project in the nearest future, which intends to support the development of the Georgian land market (DWVG 2013a).

[172] An analysis of quantities and qualities of the intended programs is beyond the scope of this work; however, with view to the fact that roughly 75–80% of the agricultural land is not formally registered at the time of programs (TI Georgia 2014a), it shall be possible to see a trend whether the condition of the support to farmers having registered their land is met.

[173] Since the 1950s, a vast body of literature and organizational structures have emerged on the issue of land administration, starting with Reports from the Land Tenure Service of the UN Food and Agriculture Organization (UNFAO) in 1953, which resulted in the Sydney Declaration of the International Federation of Surveyors (FIG 2010) with the aim to "recognise the importance of good land information and good land governance in support of the global agenda such as the Millennium Development Goals, and as a basis for meeting the key challenges of the 21st century such as climate change, natural disasters, environmental degradation, rapid urban growth, and poverty eradication" (FIG 2010: 1). An overview of the relevant literature and international organizations and their publications is provided by Williamson & Ting (2001).

However, the literature providing evidence regarding links between formalized tenure and agricultural production reveals ambiguous results.[174] Some empirical studies claim that there has been a valuable impact of titling on rural economies, e.g. through enhancing the willingness or ability of farmers to invest (Ali et al. 2011); (Galiani & Schargrodsky 2010); (Feder & Onchan 1987), triggering a rise of leasing activities (Deininger & Feder 2009), or leading to positive distributional effects (Deininger & Chamorro 2004; Schweigert 2006; Tripp 2004; Field 2005; Banerjee et al. 2002). Yet, other assessments find no significant impact due to land titling (Markussen 2008; Petracco & Pender 2009; Ghatak & Roy 2007; Boucher et al. 2005; Ouedraogó et al. 1996; Pinckney & Kimuyu 1994; Place & Hazell 1993; Place & Migot-Adholla 1989, Migot-Adholla et al. 1991). On the contrary, some authors have even suggested a number of negative consequences from fomalizing land tenure (Boone 2007; Ho & Spoor 2006; Carter & Olinto 2003; Platteau 1996; Atwood 1990, Bruce & Fortmann 1989). Such critical studies point, for example, to the marginalization of minority groups (Boone 2007), discriminatory impacts on women (Ikdahl et al. 2005) or the indirect support of influential wealthy interest groups (Deininger & Feder 2009, Ho & Spoor 2006, Carter & Olinto 2003; Platteau 1996). Additionally, there is the widely held view that alternative forms other than titling can lead to secure access to land, which in turn boosts land use, secures employment and is, hence, conducive for agricultural production (Bromley 2008b), such as planting trees (Sjaastad & Bromley 1997) based upon traditional village norms (Deininger et al; Deininger & Feder 2001; Firmin-Sellers & Sellers 1999) or by those highly endowed with social capital (Katz 2000).

"The so-called 'Washington Consensus' identified property rights reform as one of the major areas of reform for the developing world" (Trebilcock & Veel 2008: 400) and, in accordance with the mainstream neoclassical economic theory that it has relied on (see chapter 1.3), a well defined and established institutional framework vis-à-vis land has been seen as being crucial for developing market-driven economies (Cashin & McGrath 2006: 631). According to this perspective, it is assumed that formalizing tenure, if properly put in place, is a keystone of "support to economic development, social stability and environmental management" (Williamson 2001). By securing exclusive control over resources this method promises to not only facilitate land management and, hence, yield greater tax revenues but to promote and protect investments (Larsson 1991). As Bromley explains, "[e]radicating poverty is the goal, new agricultural investments, new businesses, and upgraded dwellings are the means whereby this will happen, tenure security is the necessary condition, and formal title offer security of tenure" (Bromley 2009: 20). The *individualization of tenure*, which is typically defined

[174] The following presentation of the literature is neither an attempt to offer a complete overview of empirical evidence nor a classification of findings into pro- and counter voices on formalizing tenure. It is, rather, a listing of possible impacts as claimed by the authors cited.

"as demarcation and registration of freehold title" (Barrows & Roth 1989: 1), is based on the following assumptions:

- individualization of land ownership and leasehold increases tenure security by means of reduced litigation costs;
- individualization raises investment, as a result of tenure security and reduced transaction costs;
- individualization encourages a land market to develop, because "[l]and will be transferred to those who are able to extract a higher value of product from the land as more productive users bid land away from less productive users" (ibid. 1989: 4).

In contrast, Becker (1980) emphasizes that specifically this last presumption is less supportive for justifying private property rights but, rather, serves to "justify severe limitations on specific property rights" (Becker 1980: 73), proposing that this model's dynamics are destructive vis-à-vis its fundamental assumptions of perfect competition and voluntary exchange (ibid. 1980: 73). He argues that,

> once inequality enters, a serious problem is created for the model. (…) [I]n competition for the same good, the rich will outbid the poor; and when selling the same good, will underbid the poor. (…) So in any attempt to realize the model in the world, devices designed to limit this tendency will have to be installed. (ibid.: 73)

Supported by the influential work of de Soto (2000), the strong protection of private property rights via legal property registration and the issuance of associated titles, so-called property rights *formalization*, is viewed as the key for the success of Western economies and, hence, has become "a *sine qua non* of development" (Trebilcock & Veel 2008: 401; Bromley 2008b: 20).[175] Thus, even though it has been said that

> …some of the research emanating from the World Bank in recent years has advocated a more nuanced approach to its policies relating to property rights and development, […] other documents have seemed to follow its traditional attitude that the formalization of property rights is virtually always desirable. (Trebilcock & Veel 2008: 400)

So-called "strong" property rights, meaning claims in the form of legal titles that are presumably enforced by the state, are viewed as a precondition for economic development by providing access to credit and encouraging investment in and finding the best possible use of resources (ibid. 2008: 402–408). However, it is claimed that "only when such a de jure system is in place can an economy hope to simulate further development…" (Cashin & McGrath 2006: 631). Sjaastad & Cousins (2009a: 2) point out the different understanding of de Soto and institutional economics on the role of property formalization in overall economic development, stressing that "the former sees titles as the natural and logical culmination

[175] The importance of formalizing tenure by transforming its present outward appearance from "dead capital" to "fungible assets" has been foremost supported by de Soto (2000).

of an ongoing process of rights individualisation, the latter sees state provision of titles as a necessary and desirable step in pushing the process of evolution along". The registration of (agricultural) land and other fixed assets was legally implemented in Georgia in 1998 (Gvaramia 2013: 8). Beginning in the 1990s, the activities of two donors – Germany's Gesellschaft für Technische Zusammenarbeit (GTZ), which became today's Gesellschaft für Internationale Zusammenarbeit (GIZ) as well as the United States Agency for International Development (USAID) – were highly influential in this registration process and will, thus, be highlighted here.[176]

Land management in Georgia (GTZ)

The privatization of housing space and former state-owned companies started in 1992 in Georgia (Ziegler 2005: 16): via free-of-charge distribution to actual users and persons indicated in the inventories, in the former case, and by sales and auctions in the latter. The project "land management Georgia" ("Landmangement Georgien") ran from 1994–2000 with the aim, first, to introduce a land title register and cadastre for the capital, Tbilisi; second, to set up an urban planning system; and, third, to qualify staff for administration and teaching in the areas of both the cadastre/register as well as spatial/urban planning (ibid.: 18). Due to its positive results in the setting up of a cadastre and register for Tbilisi, a further project period was fixed for 2000–2006, which was expanded to other areas and cities in the country (ibid.).[177] The GTZ-led land management project of 1996–2000 introduced a cadastre and public registry on the basis of *systematic* registration and surveyance (Laux 2005; Ziegler 2005).[178] In contrast, success in establishing an urban planning system was meager: Since the early 1990s, urban development has proceeded incidentally, determined by "individual actors and the capital strength of private investors" (ibid.: 21).[179]

Land Market Development Program I–II (USAID)

From 1997–2001, a "privatization support project" (USAID 2005: 5), namely USAID's Urban/Rural Land Privatization project, was implemented by US consulting firm Booz Allen Hamilton (BAH), which resulted in the founding of the

[176] Also other organizations, e.g. IFAD, were involved in land-reform processes (cf. (IFAD 2007). However, none of these organizations, according to various expert interviews in 2013 and 2014, have been as influential as the two mentioned above.

[177] In 2005, about 90% of the county's housing stock was privatized, with the registered land for housing space amounting to 30% of total land; estimates of the numbers of registered commercial space and differently used buildings are, however, significantly lower (Ziegler 2005: 20).

[178] The GTZ project enjoys an excellent reputation, according to several people interviewed by the author; see also Kaufmann (2003: 100).

[179] This tendency can be still observed today, according to an interview with an official of the Ministry of Economy and Sustainable Development of Georgia on August 1, 2014.

Association for the Protection of Landowners' Rights (APLR). The latter collaborated with USAID beginning in 1998 and is "acknowledged by public officials and other key informants to have been the leader in promoting completion of privatization of state-owned agricultural land" (USAID 2011b: 7). Project staff of BAH as well as specialists from APLR began the collaboration by analyzing the existing legislative framework and steering realization of a legislative basis for agricultural (and non-agricultural) land privatization and registration of titles (ibid. 2005: 6). The APLR also became in charge of…

- training programs in registration and privatization for local officials (USAID 2011b: 19), as well as legal consultation and training for the Georgian MoE (ibid. 2011b: 4);
- initial registration (ibid. 2005: 16–17): Private companies were chosen through tender to conduct field activities as subcontractors; APLR assisted in the provision of staff and technical equipment; field groups formed by one land surveyor and an assistant steered measurement activities of about 69,000 parcels on average per month (ibid. 17): "Nearly 1000–1100 people were employed in selected 32 companies. Out of this number 450 persons were qualified as surveyors and the rest represented people with different specialties" (ibid.);[180]
- operating "popular certification programs for various types of land market professionals, primarily appraisers and surveyors. The certification course for appraisers is considered the national standard in the absence of any similar certification by the state or a self-regulating professional entity" (ibid. 2011b: 4).

Thus, "[t]he pervasive influence of former APLR staff throughout the government agencies which develop land policy makes it hard to draw a line between which policies it did and did not influence" (ibid. 2011b: 10). Created in 1997…

> …as primarily an advocacy organization for newly created landowners, over this period of time APLR underwent several transitions, from advocacy organization, to donor funded technical assistance provider and project implementer, and more recently from donor funded organization to a financially self-sufficient full-purpose real property consultancy. (USAID 2011b: 2–3).

Studying APLR's history shows that "[a]long the way it has broadened its constituency to include not only the small and medium size landowners of the early years, but also large companies having vital interests in land matters" (ibid. 2011:

[180] During the LMDP, the APLR employed about 150 staff members throughout Georgia, whose number decreased to 60 employees by 2011 (USAID 2005: 17). These were hired "on a performance compensation (commission) basis" (ibid. 2011b: 4).

2–3). Though critical reports reveal "that most of its activities affected all landowners regardless of size and type of business. Its emphasis on privatization of state-owned land leased to large holders is a case in point, and those lessees are today among the largest farmers in Georgia (ibid.: 4–5).

The concepts for used registering and transacting land titles were established by the Law on Land (Immovable Property) Registration (1996) and in the Civil Code (1997), which introduced *mandatory* property registration. As many landowners were left without any proof of their landed property as a result of the initial allocation process, Certificates of Land Ownership (CLOs) were issued in 1999 by Presidential Decree No. 327 on Urgent Measures for the Initial Registration of Agricultural Land Ownership Rights and Issuance of Registration Certificates to Georgian citizens: "In order to ensure that the process of initial registration was transparent and less time-consuming, the decision was made to minimize the number of documents required for land privatization and initial registration" (ibid. 2003: 127). Additionally, Decree No. 327 "allowed the registration of the ownership of land parcels free of charge even if the area of land allocated to each household was 15 percent more than that set by the norms" (ibid. 2003: 127).

At this time, the project was handed over to USAID and transformed into the Georgia Land Market Development Program (LMDP), which was officially installed during a first phase from 2000–2005 (LMDP I) and in a second phase from 2005–2010 (LMDP II).[181] However, the role of APLR has been seen critically, considering that

> ...LMDP/APLR exerted considerable influence on policy formulation and program implementation in land privatization and registration, training scores of public officials and today's private sector market professionals. The APLR was a training ground for policy makers in government as well as for active participants in private sector property markets today. (USAID 2011b.: vii)

The objectives of the LMDP I and II were to facilitate growth, support APLR's work and provide support for developing land and real estate markets in Georgia:[182] "Activities included completion of agricultural land privatization and contributing to the establishment of a clear, transparent, streamlined and user-friendly

---

[181] With support from the ALPR, LMDP II intended to privatize state-owned forest land, which "quickly encountered opposition from within certain segments of government and the political leadership as well as a negative public and media outcry due largely to the perception that much of the interest expressed in forest land was from foreign investors, which may in fact have been true. This was perceived as a possible alienation of a beloved national patrimony and natural resource to foreigners and the government quickly withdrew its support for the concept. Today, state-owned forest land may be subject to long term concessions (up to 49 years) granted through the Ministry of the Environment, but not owned" (USAID 2011b: 24–25).

[182] The first project cycle (LMDP I) "focused on education of the public to the benefits of land privatization/ownership and the importance of formal land titles and registration" (ibid.

property rights registration system" (ibid. 2011b: 1). Besides, it was said that "titling and registration of rights would increase confidence in land acquisition, reduce transaction costs, and lead to more security for lenders and greater availability of credit" (ibid.: 19).

Decree No. 327 thus paved the way for "a *systematic* cadastre of land plots up to 2.5 ha and their registration into the Public Registry" (Gvaramia 2013: 9). The process commenced in a pilot district in Zestaponi, located in the region of Imereti (ibid.: 10–11). Likewise, (see figure 4-3 below),

> … a ceremony was held in Zestaponi on May 26, 1999, where the President of Georgia personally presented ownership certificates to the new land owners. (…) This was the first time a public event had been held for such a purpose. The presentation was attended by the US Ambassador and other officials, both from the Georgian and the US Governments. (ibid.)

Figure 4-3: Handing out Certificates of Ownership in 1999 (USAID 2005: 10)

Moreover, it is worth noting note that…

> [i]n the three years of the distribution of agricultural land to 700,000 households in the early 1990s, public opinion negatively evaluated the transfer of land into private ownership, wary of the loss of land by the farmers through sales. Dozens of seminars were held (…). Dozens of ceremonies of issuance of certificates confirming ownership on land were held (…) The ceremonies were widely covered by the press and television. Public opinion changed rapidly and drastically. In a 1995 poll of members of Parliament, 75% were against "privatization" and four years later in 1999, parliament leadership tried to organize a pro and

---

2011b: 17). The second phase (LMDP II) dealt with support for privatizing leased agricultural land, for "the necessary authority for privatization of leaseholds was not enacted until 2005" (ibid: 18).

con debate about the desirability of privatization. Not one member of Parliament would take the con position. (USAID 2005: 11)

The agricultural land privatization process, which was set in motion by Booz Allen Hamilton with 1 million parcels initially registered during 1999–2001, was handed over in 1999 to USAID and APLR and, in May 2001, to APLR and the Terra Institute (USAID 2005: 12–13). At this point, "APLR began its collaboration with Terra Institute with whom USAID had developed a Land Market Development Cooperative Agreement" (ibid 2005: 12). Thus, within the framework of the LMDP the collaboratin of APLR and the Terra Institute has been carried out in 52 districts of Georgia and has complemented initial registrations of 2.4 million land parcels (ibid.: 12). The process, marked by its rapid execution, was led mainly by political motives though, seeing that

> …the decision had been made to implement a "a quick and dirty" titling program on the theory that it was best to establish bare legal evidence of ownership for as many people as possible in as short a period of time as possible to jumpstart the market and prevent backsliding on land privatization. There was undoubtedly a political element to this decision at the time, as Georgian leadership viewed the titling program as a visible and popular initiative. (USAID 2011b: 14)

The choice for a "quick and dirty" approach has vast implications, given that in terms of the surveying work involved, "…there were few surveyors with sophisticated capabilities, there was no operational national coordinate system and no uniform standards for quality or methodology" (ibid. 2011b: 14): [183] As a result, the early groups of surveyors worked as subcontractors who…

> …were largely left to their own devices and surveys apparently were carried out by a variety of methods, ranging from use of theodolites, to chain measurements to desk review of hand-drawn maps of SDLM land arrangers made before and after mass privatization in the early 1990s. (ibid.: 14)

Moreover, " [b]ecause of the deterioration of the national coordinate system, only local or 'relative' boundary coordinates were used, and these were vaguely defined" (ibid.).[184] Thus, various types of errors occurred, "including not only geometry errors but also incorrect names, quantities, etc." (ibid.). It is estimated that by 2011 a situation had developed where "survey and other errors still arise with some frequency" (ibid.). From 1992–1999 the registration of ownership

---

[183] According to estimates, similarly conducted "quick and dirty" titling projects "can cost as much as $4–$5 per case today. Accounting for inflation, the per case expenditure for the LMDP titles may have been one-third of what would have been reasonable to produce consistently high quality mapping work" (USAID 2011b: 14).

[184] It is reported that a Law on Cartography and Geodesy was enacted in 1997 that defined a national coordinate system (UTM/WGS84) but that "it was not fully operational at that time" (ibid. 2011b: 14).

> ...consisted of a paper-based manual system maintained by the local offices of the State Department for Land Management [...] in which notations and duplicate copies of the ADR certificates issued in connection with mass privatization were kept. (USAID 2011b: 13)

The certificates obtained during the initial reform process (ADRs) were issued by a local self-governing body (*sakrebulo*) beginning in 1992 (ibid.: 8). According to USAID, the years 1998–1999 are "considered as a turning point in the recognition of agricultural land ownership in Georgia" (ibid. 2013: 6), for "owing to the assistance of international donors and primarily USAID, the process of mass issuance of land registration certificates started" (ibid.: 6) – the CLOs. In addition, USAID claimed that the registering of land was made "relatively inexpensive, simple, fast, and effective" (ibid. 2010: 7), due to the simplification and digitalization of the land registration process (ibid.: 8). Especially, before the LMDP,

> ...land registration procedures required over 60 steps, took up to two months and involved extra–legal charges. This process has been reduced to six steps, which can be completed quickly and at low cost. Land registration has also been digitalized, which has made the land market more efficient. (ibid.: 8)

From 1998–2004 the State Department for Land Management (SDLM) was affiliated to the Government of Georgia (GoG) and was in charge with

> ...land administration and management, including registering property rights and maintaining the cadastre, privatizing and leasing state owned land, categorizing and compiling land statistics, controlling the use of land and natural resources, and mediating land disputes. (Egiashvili 2013: 10)

Before that, the Bureau of Technical Inventory (BTI) "was responsible for surveying and registering apartments and buildings" (ibid. 2013: 9). Problems appeared, however, related to overlapping competencies within the Department as well as between state organizations, including the BTI, the Ministry of Agriculture and Food or the Ministry of Urbanization and Construction. In addition, political influence and conflict of interest between SDLM offices and authorities on the central and local level led to corrupt practices and an ensuing underfunding of the organization (Egiashvili 2012: 1–2). Beside this, the BTI was reported as being "an institution that routinely extorted property owners who requested the necessary documentation for registration, but which also imposed an extra, time-consuming step on registrants" (USAID 2011b: 5). In the years 2004–2006, the data was transferred from SDLM to NAPR by "using common mapping standards and technology. (...) Ultimately LMDP assisted NAPR to transfer the integrated database to the local offices of NAPR, including system consultation and training for local technicians" (ibid. 2011b: 6–7). The electronic cadastre was initiated during 2006–2007 (ibid.: 21). In 2008, the Law of Georgia on Public Registry was enacted (which replaced the Law on Privatization from 2005). The NAPR, is says, is "a body operating under the Ministry of Justice of Georgia and responsible for maintaining a public registry" (Art. 1.1). More specifically, it is "a semi-independent state company that reports to the GOG through the Ministry of Justice" (USAID 2011b: 5) and is, since 2009, in charge of the physical cadastre as well as the registration of property rights to immovable property, "registration of legal entities, and registration of secured transactions with moveable property" (ibid. 2011b: 20). Unlike other aspects of this process which have been subjected to criticism, "[t]he registration function of the NAPR is rated second in the world in the World Bank Group's 'Doing Business' assessments and enjoys an excellent reputation for public service among system users" (ibid.: 5). Due to this process, the following bodies became in charge of land management and land administration (Egiashvili 2013: 10):[185] The Ministry of Agriculture (MoA) in charge of land management, land use planning, monitoring, and alienation of land; Ministry of Economy and Sustainable Development (MoE) responsible for urban land planning and the selling of both agricultural and non-agricultural state land; Ministry of Energy and Natural Resources in charge of forest management; the Ministry of Environment Protection in charge with the management of national parks and protected areas; as well as local self-government bodies (*sakrebulo*) that are in charge of the management of non-agricultural lands as well as planning in coordination with the MoE.

At the end of the project, the following target groups were identified as being the most vulnerable to the new legislation (USAID 2011b: 22):

- Land users who were not provided with a certificate of land ownership during the first land distribution reform (ARDs) or at the end of the 1999s (LMDP's CLOs): "Such people are thought to exist, but there are no good estimates of numbers" (ibid. 2011b: 22);
- Farmers "who have received land rights from the local land or "Acknowledgement" commissions since the completion of CLO titling, without registering (ibid.: 22): "There is some evidence that between 2002–2008 a number (unknown) of people continued to receive land rights from the local commissions and that these people may not hold CLOs or may not have registered" (ibid.: 22); and
- Other farmers "whose boundary surveys from the CLO period are erroneous and who have not yet prepared a new survey for entry into the electronic cadastre" (ibid.). The latter is thought to give rise to the most prevalent risk, for "[i]n some of those cases land occupied and used by the citizen may be characterized as state-owned and subject to further sale" (ibid.).

Multi-donor supported formalization

[185] A detailed overview of the relevant bodies is provided by Gvaramia (2013: 21–22).

At the end of the 1990s, not only GTZ (today Gesellschaft für International Zusammenarbeit, GIZ) and USAID were involved in the reform but a number of other international organizations were also assisting this process (see table 4.1 below). The eight donors involved in land administration projects included the International Fund for Agricultural Development (IFAD), World Bank, USAID, KfW Development Bank (KfW), United Nations Development Program (UNDP), European Union (EU), Swedish International Development Cooperation Agency (SIDA), and GTZ (IFAD 2007: 23; see table 4-1 below).

Donor	Project	Aim
Deutsche Gesellschaft für Technische Zusammenarbeit (GTZ)	Urban Land Management Project (1994–2007)	Funding of cadastral surveying and registration activities in Tbilisi; institutional support for land administration and land management.
World Bank, Fund for Agricultural Development (IFAD)	Agricultural Development Project (1997–2005)	Funding of basic aerial photographing, mapping, cadastral surveying and registration, and the refurbishment of offices in two districts. The second phase of the project envisages the establishment of regional cadastre centers.
US Agency for International Development (USAID)	Land Market Development Project (1997–2012)	Helped privatized agricultural land mapping, surveys, registration and support for issuance of ownership certificates in 51 districts. Support for land ownership enhancement and increased public awareness in Georgia.
United Nations Development Programme (UNDP) and European Union (EU)	Land Management Project (1998–2004)	Funding of ortho-photo development, cadastral surveying and registration in one district, software development and the refurbishment of 11 district offices. The EU covered the government's contribution under its Food Security Programme until 2000.
German KfW Development Bank (KfW)	Cadastre and Land Register Project (2000–2008)	Funding of base aerial photographing, ortho-photo development, mapping, cadastral surveys and the establishment of six regional centers in the developed territories of Georgia.
Swedish International Development Cooperation Agency (SIDA)	Capacity Building & Improved Client Services at the National Agency of Public Registry in Georgia (2001–2004, 2008–2013)	Funding of capacity building by establishing a training center for SDLM employees on land legislation, land registration, land information system, cadastral surveys, valuation and taxation, credit marketing and office management. Improvement of NAPR strategic management development, improvement of quality of services, etc.

Table 4-1: Multi-donor support for land registration in Georgia, 1994–2013 (adapted from Sa lukvadze 2006: 9)

But the first stage of implementation (1996–2002) was marked by insufficient coordination between the donor projects, "with some portion of noticeable rivalry. These circumstances were caused by absence of a common concept of system design and shared vision on final results and products" (Salukvadze 2006: 19). Thus, "each particular project followed divergent aims, implementing own approaches and methods, covered areas and operation segments of its own interest" (ibid. 2006: 19). An advanced GPS surveying reference system was "financed partially by the World Bank" (ibid. 2011b: 4). The application of different technical approaches and facilities used by the different donors (see figure 4-4 below) is exemplified by the land surveying techniques employed by USAID and KfW from 1999–2002. The USAID project targeted privatized land parcels from the so-called reform fund and arranged two million ownership certificates by "using traditional survey instruments (old theodolites, tapes, etc.) fixing parcel boundaries in local/relative coordinates, thus, leaving them non-georeferenced…" (ibid.: 18).

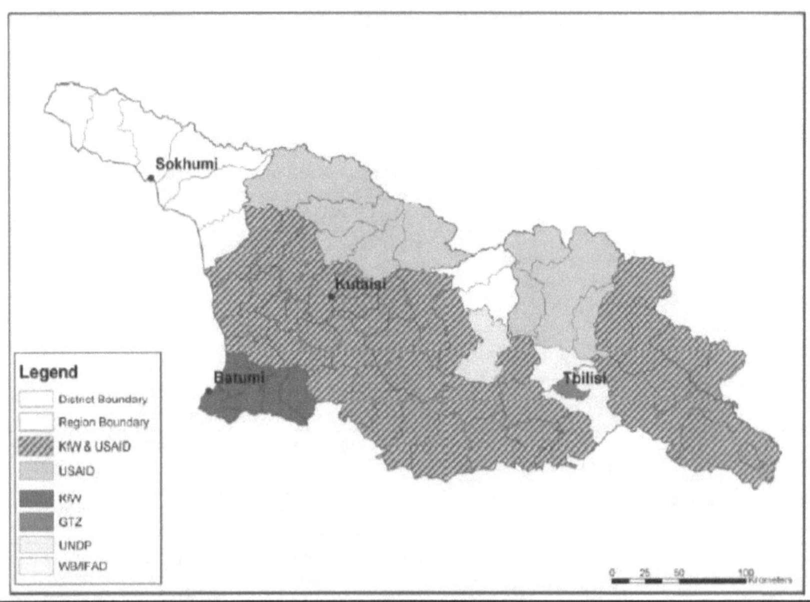

Figure 4-4: Area of donor-led land administration projects (Salukvadze 2006: 9)

Meanwhile,

> …after 2–3 years KfW project measured remaining neighbouring parcels – households, leased and state owned land (almost 1.5 million in total) – using high-accuracy orthophoto background in combination with GPS based modern digital plane tables (DPT) and fixing boundaries in UTM co-ordinate system. (ibid.: 18)

Compilation of data and harmonization of the technical systems to one cadastral database at NAPR were carried out by the German KfW Development Bank during the second implementation phase of 2004–2012 (Salukvadze 2006: 9).

According to Egiashvili (2013: 11), the year 2004 marks a change where "[s]urveying services were shifted to the private sector, thus creating a competitive environment for surveyors who can be held liable for the quality and reliability of their work". The survey work registered by NAPR, however, was marked by inaccuracy, as "survey and other errors still arise with some frequency. The extent of the undiscovered and uncorrected errors is not known, and estimates of current registrars range 10% to over 50% of cases…" (USAID 2011b: 14).

In 2007, the Government legally amended the formerly designated *systematic* registration process toward a *sporadic* registration process (Law of Georgia on

Recognising Property Rights Under the Possession [Ownership] of Physical and Private Legal Entities [2007], Art. 4.2): "The recognition of the property right to lawfully possessed (used) land shall be exercised through registering the right of a lawful possessor (user) in the Registry". The Law also stipulates that, from 2012 on, legal persons "…shall lose the right to have the property right to lawfully possessed (used), as well as to squatted land recognised. After the above time limit, the property right may be acquired according to the general procedure determined for privatising state property" (Art. 7.4). As about 30% of the newly created land-owners have not obtained any proof of ownership, the Law provides for applying to an archive of the relevant municipality keeping former records (Gvaramia 2013: 8–9).[186] In cases of lawful possession, the following steps are required (ibid.):

1. Provision of a notice regarding the land from the archives (estimated to GEL 42 [USD 17]);

2. The application must "[h]ave an electronic draft survey of the measurements of the land performed (GEL 60 [~ USD 24] – minimum payment for land up to 500 square meters)" (ibid.);

3. Paying a registration fee for "the cost of service from the public registry and bank (GEL 51 [~ USD 21] non-refundable)".

Thus, for an applicant holding land in lawful possession "the cost comes to a minimum of GEL 153 [~ USD 62] for registering one piece of land up to 500 square meters" (ibid.). TI Georgia hence criticized these costs in relation to the "low income of the local inhabitants" (ibid.), especially in cases where people own three or four land plots, "which requires them to pay the registration costs for each piece of land that they own. Registering land held under unauthorized occupation may lead to higher costs, depending on the location of the land." (ibid.). As a consequence,

> …land sales often take place solely on the basis of a transfer certificate (…) and the transaction does not get officially recorded within the land register. Such informal transfers of property ultimately will complicate proof of ownership and facilitate respective disputes, especially over the long term. (KfW 2011: 4)

[186] Municipal archives hold at least one of the following records, namely certificates issued by the Technical Bureau Archive of the Public Registry, an excerpt from the Household Book, the list of land distribution, the book of the gardener or a certification on assets from a district archive (Gvaramia 2013: 8–9).

Further legislative changes in June 2012 initiated free-of-charge "primary registration of property rights on land as well as registration for specifying the area" (ibid. 2013: 9).[187]

Regarding pasture land whose governance was re-transferred from sakrebulos back to the central government in 2010, two problems arose. On the one hand, the Georgian legislation does not recognize communal land ownership, thus almost 1,800,000 ha of pasture land used by community groups "remains unregistered and the land rights of these community groups are not protected" (Egiashvili 2013: 7). On the other hand, a number of pastures were registered as municipal property into the Public Registry when they were under municipal ownership (2006–2010), "[e]*ven though the legislation explicitly states that the Ministry of Economy and Sustainable Development of Georgia is the owner and manager of pastures*" (Gvaramia 2013: 19; italics in the original).

In a third report by TI Georgia, "Stripped Property Rights in Georgia", the non-governmental organization lists further general problems with regard to the new registration process (TI Georgia 2012b):

- threats of property violation, especially in touristic zones: 271 residents in Gonio (Adjara) were deprived of property by the federal government; infringement of ownership rights of local residents in Mestia (Svaneti), "and the difficulties the local population encountered when registering their titles to land plots" (ibid. 2012c: 3);
- cases of abandonment, where "citizens give property to the state as a gift" (ibid.: 3).[188] The newly established NAPR "is unable to compare drawings developed through the application of two different systems (hard and electronic versions of cadastre drawings) allowed under Georgian law" (ibid.). Thus, "it can register anybody's title to property already registered in somebody else's ownership without identifying a conflict or overlap between the submitted applications. Unfortunately, the courts often fail to protect the victims" (ibid.). As a result, TI Georgia (ibid.) reports that "…landowners suddenly discover that their land has become someone else's property" (ibid.)

Chronologically, TI characterizes violation of people's property rights after the Rose Revolution in 2003 in different waves, where the first infringements

> …occurred at the end of 2003 and beginning of 2004. Subsequent waves were more geographically concentrated and related to the municipal organization of individual towns

[187] The role of primary free-of-charge registrations was a tool used by the Georgian Government and not by those wanting to register their property in-use, as the qualitative study discussed below shows (see chapter 5.2.2).

[188] TI Georgia (2012b: 3) informs specifically that "…twenty cases of giving property to the state as a gift and two cases of abandonment of property were reported between December 1323, 2010, in Sairme. 79 facts of abandonment of real property by private persons were reported between January 13 25, 2011, in the territory of Bakhmaro resort".

> or regions, like in the cases of Sighnaghi (2007) and Rike in Tbilisi (2007), when private owners gave up their property as a gift to the state. In these specific cases it was peculiar that such valuable property (and a significant source of income) was given to the state as a gift, especially when the state's intention to develop the land where this property was located had already been announced. (ibid.: 4)

Moreover, further critique has been targeted toward the critical role of land commissions, which have been acting since 2007 as local commissions in charge of recognizing lawful or unlawful possession of agricultural land (see above): "These commissions are functioning in ways that put people's rights at risk or prevent them from obtaining rights" (Rolfes, Jr., Leonard & Grout 2013: 1). A number of court cases reveal "inconsistencies in decision-making and adherence to the law to raise doubts about the impartiality of the judgements being rendered" (ibid. 2013: 1). Another critique points to the neglected role of women in the first years of the reform: although land plots were initially designated to households, the rights of spouses were neither legally recognized nor registered (ibid.: 1). Furthermore, pasture land has in some cases been registered by municipalities, although the MoE is the actual owner and in charge of managing pastures. Moreover, according to another report by TI Georgia, property registration has met hurdles in the form of "no-registration zones", which were established for example by the Public Registry Registration Office of Mestia in 2011 (TI Georgia 2012a), where

> …requests to register property located in these zones have been denied, even though there is no legal basis for denying registration for this reason. […] On May 3rd, 2011, at the request of the Ministry of Economics and Sustainable Development, the Public Registry of Mestia registered the ownership of […] land located in a no-registration zone […], where, according to local journalists, the building of the presidential residence is now underway. (ibid.)

It also turns out that problems in relation to the registering of land in the no-registration zones of Mestia were resolved "only in cases where there are specific plans for building structures or implementing infrastructural projects and the investors agreed to compensate the landowner" (ibid.). Otherwise, "registering ownership of land in these zones has not been possible to date" (ibid.). As a consequence of these "[o]ften (…) prolonged, difficult and costly processes, local inhabitants might suddenly find out that the land they have traditionally owned for centuries has become the property of the government or some other private individual" (ibid.).

4.5 Summary

Georgia's transition from a planned to a market economy was a drastic and far-reaching process, from which its populace has yet to recover. Although agricultural land has always been a scarce resource, Georgian agriculture was vital for the Soviet production and distribution system (e.g. wine, tea, citrus fruits). Nowadays, however, the sector only plays a minor role in the overall Georgian economy, with the common household plot of about 0.25 ha still being central for subsistence. Georgia's first years of independence were marked by inter-ethnic conflicts, civil war and the deterioration of the economy, due to the break-up of the interdependent Soviet production and distribution system. In need of financial means, the government followed a pro-Western course that led to a drastic rise in foreign debt, foreign assistance and humanitarian aid. Restitution was avoided, seeing that Georgian society had still been organized feudally before becoming a Soviet Republic. Thus, land distribution was chosen to avoid famine and enable subsistence production, resulting in the distribution of two or three land plots of 1.25 ha on average to rural residents, based on local availability not land quality.

The introduction of private ownership, though, has often been underestimated in scale and scope. By 1998, about 60% of the overall agrarian land had been privatized, with about 15% more land per household having been distributed to households as foreseen. Yet, a number of cases revealed discrimination against ethnic minorities in receiving land. Furthermore, land allocation has often lacked any legal guarantee. Although a law from 1996 stipulated that private land ownership is a privilege of Georgian citizens and legal entities registered in Georgia, the process was often hindered or even blocked by local elites via bureaucratic hurdles. Meanwhile, in the further course of the reform process, the vast majority of rural dwellers could not afford the required payment to obtain necessary legal documentation regarding ownership (the so-called Receive-and-Delivery Act).

Beginning in 1996, the government initiated a market-oriented reform with the aim of leasing out agricultural land remaining under state ownership for 49 years. But the process has been characterized as biased and corrupted by influential government officials, with most of the land being held for economic (sub-renting) or speculative purposes, as data and a number of cases reveal. Newly enacted legal regulations by the new government in 2005 aimed to privatize leased land as well as unused, vacant land. It provided leaseholders with preemptive rights to acquire ownership rights from the state within a five-year period, together with preferential payment schemes over a ten-year term. Lands that were not acquired by lessees were either auctioned firstly to local, ethnic Georgian legal subjects or, otherwise, auctioned to any kind of bidder. The process was marked by fraud, resulting from financial and organizational hurdles that limited and complicated the process for lessees to acquire ownership rights, according to reports about the newly created land owners. In 2010, the same government retransferred communal ownership rights of pasture land back to the state and opened up a window for the acquisition of pastures by setting regulations with a retroactive effect: Pastures leased for more than five years became subject to privatization and had to be acquired within about a year. The remaining pasture land was leased out or auctioned by the Ministry of Economy and Sustainable Development, later in charge of land privatization. Further efforts to attract private investors and promote their engagement in agricultural ventures seem to have been disorganized and limited in scope. Consequently, the final outcome of these efforts, namely how much land is now under state ownership, is not known today. Popular initiatives for the acquisition of land by (a small number of) South African Boer and Punjabi farmers was met by resistance from the local population, who eventually filed a petition to stop the selling of land to foreigners. Not being in line with the Georgian Constitution, such endeavors were thus suspended in 2014.

Reforms to realize the formalization of land ownership from 1996–1999 failed due to outdated data, as well as insufficient organizational and financial means. Under the influence of initially German, then American and later multiple donors, formalization of ownership was promoted by massive campaigning for individual, private ownership and, by the end of the 1990s, the systematic registration of ownership titles together with the issuance of legal documentation – Certificates of Land Ownership (CLOs) – which was free-of-charge in some areas of Georgia. But the methods and strategies of the various donor organizations were marked by competition. Each project applied different technologies and surveying activities, which produced incompatible data that needed to be harmonized for use in the newly established digital land management system for property registrations and cadastre of the National Agency for Property Registrations (NAPR). The latter commenced registration in 2004 and launched the cadastre in 2006. Before that, several state bodies had been in charge of land management and land administration, which led to overlapping of competencies; moreover, several cases revealed violation of people's right to register their property. The decision to employ a systematic registration process was altered in 2007 toward sporadic registration. Meanwhile, a number of cases revealed violation of people's registered property rights, including evictions of farmers from their individually held land, especially in potential touristic zones, and the revocation of communal use rights or infringements to rights to formerly leased pasture lands. Inaccuracies in the "quick and dirty" land-surveying measurements taken slowed down the process of completing the cadastre. Various bureaucratic obstacles also hindered local inhabitants from registering property, as has been revealed by court cases that were marked by inconsistencies in decision-making and adherence to the law. The costs of registering property have also been estimated as being relatively high, but legal changes in 2012 led to initial registrations being granted free-of-charge. Though costs were drastically reduced and the necessary steps for registration of property have been simplified, registration rates have remained low, with only about a third of the land in Georgia having been registered by now. The following analysis seeks to show who benefited and who lost due to these reforms, while also shedding light on the role that land-tenure formalization is playing in agricultural production in Georgia today.

5 FINDINGS

A commonplace of experience is that decisions nominally made by governments often register determinations that are made outside government – whether in a bishop's place, a club of industrialists, or a trades-union headquarters. More generally, in many sectors of human life the norms of conduct are formulated and made effective outside the machinery of legislation, administration, and adjudication.
(Lasswell 1971: 2)

The following chapter seeks to shed light on the actual outcomes of the Georgian reform process. In the first part, the results of a survey by the Center for Social Studies (CSS) are quantitatively analyzed via SPSS to help in characterizing the resource users who are investigated in this study, namely small- and large-scale agricultural producers and processors of hazelnuts and wine grapes. Moreover, the survey analysis helps us to understand their organization of everyday agricultural production and focus on factors that affect registration of agricultural land in Georgia as a final requirement for obtaining private ownership of it. The results show that financial means are decisive for making the choice to register property. Additionally, they indicate that villagers prefer agricultural production internalized within the family rather than achieving potentially higher income in agriculture by cooperating with others; being involved in the village community seems to substitute for seeking the sanctuary of so-called secured land titles. The method used for obtaining these results and a discussion of further findings is the subject of the subsequent subsection (5.1). The second part of the chapter is concerned with a qualitative analysis of key features of the privatization process (5.2). On the basis of Grounded Theory and facilitated by the Atlas.ti software (see chapter 2.1), I then analyze the impacts of the reform legislation (collective-choice rules) on the prior rules-in-use (operational rules) vis-à-vis land privatization and agricultural production in order to evaluate the outcomes of the reform process.

5.1 Quantitative empirical findings

After data was collected from the sample population on the basis of a questionnaire CSS carried out in 2012, it was entered into an SPSS database. However, before analysis commenced, tentative working hypotheses were developed regarding the assumed relationship between the choice to register (dependent variable) and socio-economic factors (independent variables). By comparing anticipated relationships with those actually found on the ground, the purpose underlying this initial step was to make explicit what was implicitly presumed and see if the empirical data either verified or disapproved my initial assumptions. For example, taking the first question of the questionnaire, "[c]omparing the economic situation of your village to other villages in the region, to which of the categories listed below would it fit into?" (A1), I had presumed that villages classified as "rich" would exhibit higher registration rates than those classified as "poor". However, the analysis demonstrated that this assumption was mistaken, as the economic condition of a village does not appear to be an indication for having higher registration rates (see variable A1, chapter 5.2.2).

The analysis was organized in the following way: The data set was, first, calculated according to a response's frequency distribution, whether it was valid or invalid (N) and, if relevant, its mean (and median), standard deviation and sum. It was then possible to reveal deviant distributions and contrast these findings more easily with other sources of information. Second, based on the dichotomous choice of the respondents about whether "the land [is] registered in the public registry" (F5), two groups were formed – registered vs. not registered – and compared in the subsequent steps. The figures of each group were then entered manually into Excel to produce graphs illustrating the respective relationships of the variables in question. In cases where striking connections were found, the variables were tested for significance.[207]

The results were grouped according to the following categories: (5.1.1) General picture of the formalization of property rights, including spatial and demographic characteristics; (5.1.2) household and farm characteristics; and (5.1.3) village cohesion and collective and informal rules, especially those illustrating institutional changes the village members had perceived.

5.1.1 General picture on tenure formalization

When asked "how many hectares of land does your family own?" (F4), almost all of the interviewees (only four out of 600 refused to answer or did not know) responded, with an average (median) amount of land in their possession being about 0.8 ha per household (standard deviation of 1.03).[209] This figure is roughly equivalent to the number of about 0.75 ha usually mentioned in the literature (Lerman 2005: 1). Considering that earlier findings mostly date from about ten years prior to the CSS survey, these first results seem to indicate that land distribution remained relatively unchanged over the years.

According to CSS' numbers on the formalization of property rights in agricultural land (F5), less than half of the respondents (N=596), namely 43.3%, stated to have registered their land in 2012, while 51.7% did not. These figures contrast sharply against those projected by KfW and the World Bank from 2011, which

[207] The null-hypothesis was tested and rejected at a significance level of $p=<0.05$.

[209] The modus accounts for 1 ha/household; half of the respondents (50.7%) possessed ≤0.8 ha; only 14 households (out of 596 total) possessed more than 2 ha land, and only four households held eight or more hectares of land in their possession, namely 8, 10, 10.8 and 15 ha.

estimated the share of those who registered or not to be 9% and 16.8% respectively (KfW 2011: 9). Seeing that this growth denotes a quite drastic increase within a year, the number should be treated with the utmost care.[210] The results for the year of registration (F6) illustrate a steady increase within the decade prior to the survey, displaying a massive rise in 2006 and again from 2010–2011 (see figure 5-1).

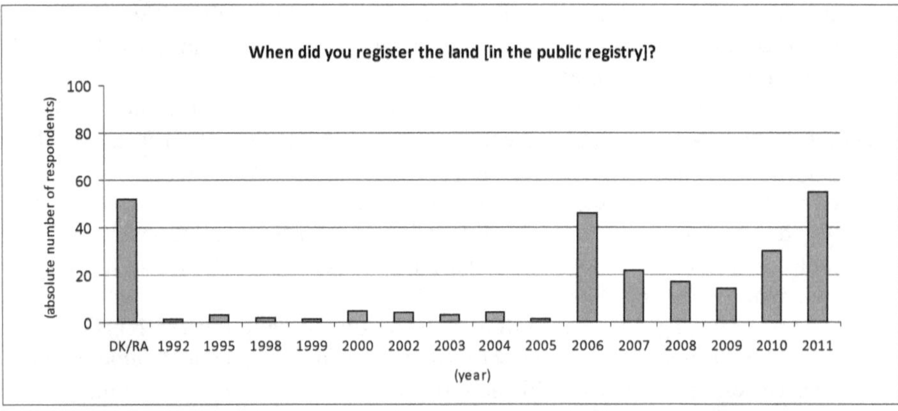

Figure 5-1: Land registration in Georgia by year, according to CSS survey (2012) respondents

A new state act, enacted in 2005, granted lessees and land users preferential rights for land acquisition until May 2011.[211] The decline of registrations in the meantime may be explained by the deteriorating relationship between Georgia and Russia, expressed by the Russian ban on Georgian agricultural imports (2006–2013) and a creeping militarization that ended in the so-called August War with Russia in 2008 (Tsereteli 2006; Halbach 2006, 2007, 2008). A share of 20% of those who registered their land but who responded that they "don't know/refuse to answer" seems to indicate that this information was not very present in their minds.

Of those respondents who did not register their land (n=312), 64% claimed to have no money or time to register (F7). Despite the fact that this possible answer, as pre-formulated by CSS' questionnaire, included two different aspects – time and money – the dilemma regarding which of the two might be the real cause may be resolved by looking at the responses to a further question, focused on the involvement of household members in agricultural work (F3): More than half (56%)

[210] The situation regarding precise land-amount figures is further aggravated by the fact that people usually possess about three or four plots of agricultural land (Lerman 2005: 1). It is, hence, not clear whether the responses by interviewees of having registered their land refers to all of their plots or just to one or a few. The problem has been dealt with during my last research visit in 2014 but could not be settled entirely.

[211] The Law on Privatization of State-Owned Agricultural Land of 2005 was suspended and replaced by the Law on State Property of 2010 (USAID 2010).

of those who did not register land declared that "almost all" worked, in addition to another third (33%) who indicated that "several members of the family" were involved in agriculture. Hence, it makes sense to assume that some household member would have been available to work in the absence of the family head during the time required to register their land. Consequently, I assume that the reason for not registering land was money, not time. A further reason for not registering was lack of information, as stated by 22% of the respondents. A minor share (5%) described that their "documents [were] not in order", while a few (2%) would have registered their land if they had sold it.

Registration numbers diverge considerably, both from village to village (sof_1) and also within the same region (N=565) (see figure 5-2 below). For example, in one village of Kakheti in the East of Georgia, the vast majority did not register their land (76.7%; in contrast to 13% of those who did); meanwhile, opposite results were found in another village of the same area, where the majority (70%) registered land and 26.7% did not. Registration behavior in Samegrelo in Western Georgia exhibits equal results overall but not less divergent. In one village, the majority (76.7%) did not register, in contrast to 23.3% who did, while in another village the share of those who did not register (43%) is slightly outbalanced against 40% who registered; a third village is again dominated by those who registered (53.3%), in contrast to 36.7% who did not. Thus, we can say that the data shows that registration rates were unevenly distributed among and within regions.[212]

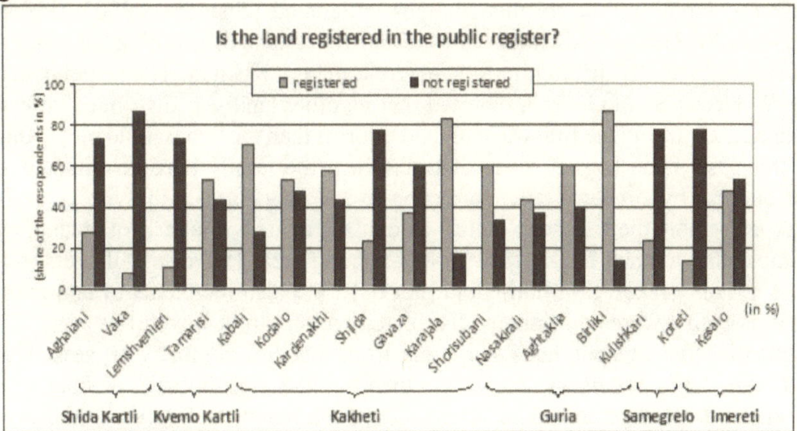

Figure 5-2: Land registration by geographical area in selected regions of Georgia, according to CSS survey (2012)

[212] Whether the location of a respondent's village, i.e. situated nearby major roads or hard to access, is relevant for their choice to register land or not could not be analyzed within the framework of this survey; the question was therefore addressed during the qualitative research (see 5.3). However, as the analysis will show, neither proximity to major roads nor to the village proves to be a clear-cut indicator for or against registration.

5.1.2 Household and farm characteristics

The typical household is comprised of four household members (F1) and, as has been mentioned above, possesses about 0.8 ha of land (F4). Generally, among more than half of the population, all (54.3%) or at least several (35.8%) family members were involved in agricultural work; only a small share (6.8%) of household heads worked alone (F3).[213] With such a large share of family members involved in agriculture it is no surprise that the largest group, 67.5% of the respondents (N=600), mentioned working on the "[l]and plot at my house and all other land plots" (F9) owned by the family. The second largest group (22%) worked "[o]nly the land plot at my house", while merely 6.2% of the respondents worked on the "[l]and plot at my house and over half of other land plots".

How can this divergence be explained, namely that the majority of respondents worked all of their land, while the next largest group only worked on their household plot? This difference might be explained by the circumstance that, after the initial land distribution, the vast majority of the people in Kakheti were able to acquire additional land, but the people of Samegrelo were not. However, seeing that agricultural productivity declined drastically throughout Georgia after independence (WTO 2009a), it is not convincing to imagine that Kakheti villagers found financial sources for the acquisition of further ground, while those from Samegrelo did not. Thus, there must be another feasible explanation.

Although the literature mentions distribution of a household plot and further pasture land up to 5 ha for mountainous settlements (Ebanoidze 2003; Bezemer & Davis 2003), my fieldwork in 2014 revealed that land distribution has also been carried out slightly differently in Samegrelo than in Kakheti. The respondents in the Western part of Georgia reported that no other land was distributed after independence during the first wave of land reforms than the "household plot" which, in this case, is a "larger" piece of land where the family house is situated and surrounded by orchards and gardening areas. In contrast, in East Georgia's Kakheti region, the "household plot" amounts to a much smaller plot of land, with a house and a small backyard with a gardening area. However, villagers in the East not only received a household plot but were also recipients of agricultural land plots that were additionally distributed with the first wave of reforms. Consequently, those respondents stemming from Samegrelo mostly possessed, and thus can only work on, no other land than the plot their house was situated on, whereas respondents in Kakheti came into possession of their household plot and additional land (see 5.2). This assumption is supported in my quantitative analysis by the outcome for the village Kulishkari, where 93.3% of the respondents stated to work solely on their household plot (unfortunately, data of two further villages in Samegrelo, Akhalsopeli and Guripuli, do not exist for variable F9). But has land been distributed differently in Samegrelo only, or is that a phenomenon spanning over the entire West of Georgia? This question prompted further analysis by

[213] The question includes children from the age of seven (CSS 2012).

forming two sub-groups according to two broadly defined regions differentiated by their location: one group called "East", comprising the villages in Shida Kartli, Kvemo Kartli and Kakheti, and another grouping called "West", containing villages in Guria, Imereti and Samegrelo.[214]

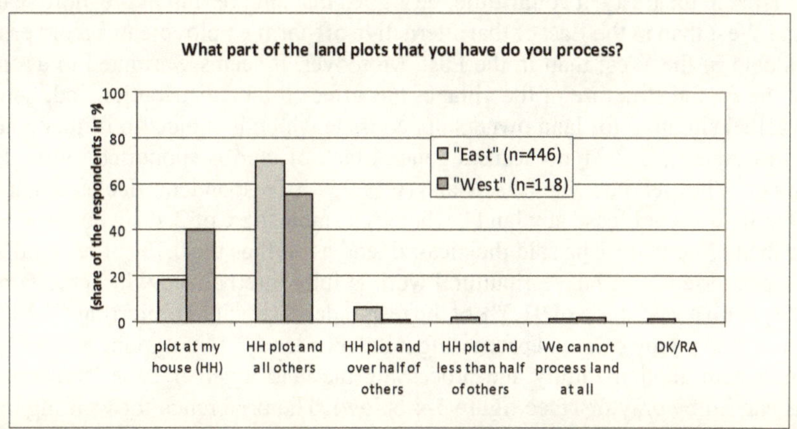

Figure 5-3: Type of land plots worked by family members, according to CSS survey (2012)

A contrasting juxtaposition illustrates a significant relationship between a household's location and land area worked (see figure 5-3 above).[215] As anticipated, the majority (55.8%) of the "West" group worked their household plot plus all other land owned, though another considerable share of Western respondents (40%) claimed to work only their household plot, and a minor share of 1.7% reported not working any land at all. Meanwhile, the distribution of responses from the "East" grouping is as follows: Here, the major share (70.2%) worked their household plot and all other land owned, while just a small part of the sample population (18.4%) only worked their household plot, an even smaller share worked their household plot plus more than half of the remaining land (6.4%), and another worked their household plot plus less than half of the remaining land (1.6%); only 1.1% of the Eastern-located respondents did not work any land at all.

Seeing that a considerable share of the Western inhabitants also worked more land than just their household plot, it seems wrong to argue that people in the West were recipients of no more land than one household plot during the first wave of

---

[214] As mentioned in 4.2, within the CSS (2012) data set, the group of Easterners is heavily over-presented, with 450 respondents – representing three-fourths of the sample population (75%) – in contrast to 120 respondents originating from the West (20%); meanwhile, 5% of the respondents are somehow missing from the total.

[215] According to a Chi-square test, the null-hypothesis is rejected significantly for both groups (p=0.000).

land reforms. Nevertheless, the figures show a clear correlation between those families located in the West who were more likely to possess and, hence, worked only the plot around their house and those originating from the East, who generally worked their household plot plus additional land. Considering the relatively small number of respondents who did not work any land at all – 1.7% in the West and 1.1% in the East – it could either be argued that land resources are more scarce in the West than in the East or that alternative off-farm employment is more easily available in the West than in the East. Moreover, it seems warranted to assume that the spatial structure of the villages has affected land distribution and, hence, the actual situation for land ownership, an issue which is subject to further analysis in chapter 5.2. A hypothesized general lack of land is supported by the low number of households who leased land (F9): 95% of respondents declared that the "[f]amily does not lease any land", whereas a small share of 2.7% who possessed less than or equal to 1 ha said they leased land as well as the 0.7% of respondents who possessed 2–11 ha. Agricultural work is fully internalized within the family (B24), with a vast share of 91.2% of the respondents (N=565) preferring "[h]aving 2 ha of land on my own and processing it independently" to "[h]aving 6 ha of land shared with another family and processing the land together"; the latter option was chosen by only 6% (see figure 5-4 below). The preference for working individually, within the family (B16), is underlined by more than half of the respondents "rather" (38.3%) or even "fully disagree[ing]" (21%) with the opinion "that if several families unite in the village to conduct their agricultural work together, they will live better"; only a third, namely 7.7% "fully" and 23% "rather fully agree", see working together as the better option.

The relationship between registration rate and preferences regarding individual work is not significant, but shows a trend that those who agree to unite with other families and work jointly in agriculture have registered less.[216] However, the figure indicates that the vast majority of respondents prefers to work independently.

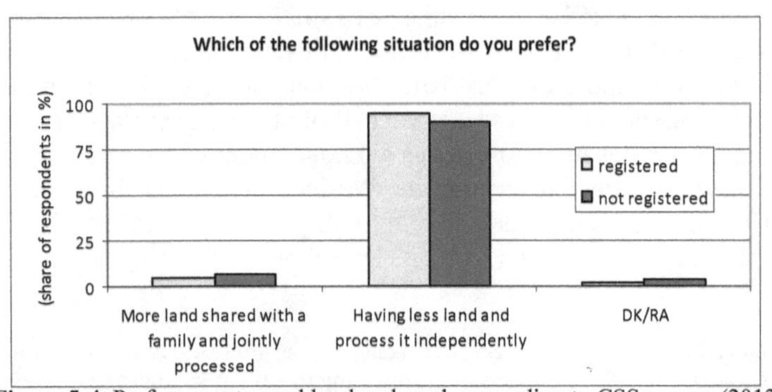

Figure 5-4: Preferences toward land and work, according to CSS survey (2012)

[216] The t-test for this interval scale was not significant; a Chi-Square of 0.65 shows a clear trend.

Almost a third (29.5%) of the sample population (N=600) owned one cow (F10), 13.7% possessed two, and 11.3% three cattle. Overall, a total (cumulative) share of 69% owned up to ten of these animals, while 5.7% owned up to 40 cattle.[217] 25.3% of the respondents did not keep any cattle at all. In monetary terms (F19), two-thirds of the population estimated their monthly income to be between GEL 101–200 [USD ~ 41–81] (29.8%) and GEL 201–300 [USD ~ 81–121] (31.7%), while 16.2% of the households received GEL 301–500 [USD ~ 122–202] per month and 3.7% earned more than GEL 501 [USD ~ 203]. It is not surprising that the correlation of income levels and registration rates proved to be significant: those with higher income tended more often to register their land.[218] In terms of the adequacy of their incomes (F14), more than half (54.7%) of the respondents stated that they bought "food and items of primary consumption" and 26.8% acquired "clothes and minor expenses", whereas 16.7% did not have enough food. In contrast, 1.3% stated that "[f]rom time to time we manage to make big purchases" and 0.3% of the respondents said that their "[i]ncome is enough for everything, we manage to save up". The figures on nationality (D3) show that, with 74.7% of the sample population, ethnic Georgian respondents were much more represented than their Armenian (21.2%) or Azeri (4%) counterparts (see figure 5-5 below).

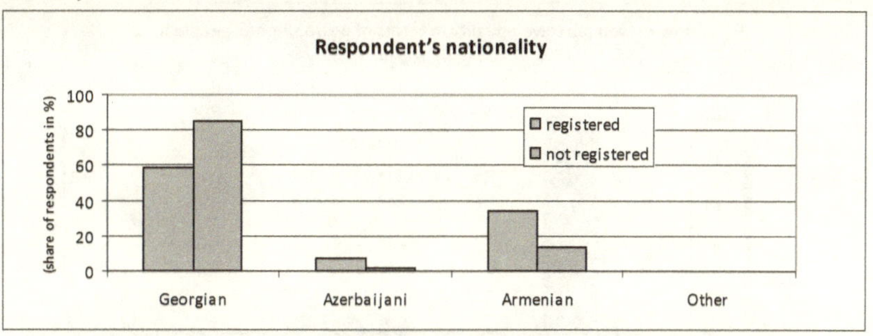

Figure 5-5: Respondents' nationality, according to CSS survey (2012)

Even so, the analysis clearly shows that both minorities of Armenian and Azeri origin tended to register more than their Georgian fellows (D4).[219] The respondents' religion exhibits a trend of correlation with registration rates, with Orthodox Christians registering less often than Muslim respondents.[220] These findings seem

[217] Surprisingly, a remarkable share of 25.3% of the respondents did not know the number of animals they possessed or refused to answer.

[218] The t-test done for this interval scale indicated a probability of p=0.000 and, thus, proved to be highly significant.

[219] A Chi-Square of p=0.000 points to a clearly significant correlation between nationality and registration (nominal scale).

[220] The chi-square test used for this nominal scale revealed a probability of p=0.087 (D4) and, hence, reveals a trend of interdependency between religion and rate of registration.

astonishing, given that empirical accounts from other settings, usually in African contexts, commonly report the contrary due to discrimination against minorities (Boone 2007: 585), and thus will be treated further in chapter 5.2.

5.1.3 Cohesion and collective action in villages

The economic level of the respondent's village (A1) was generally perceived as being "medium" by 63% (N=600), by 27.8% as being "poor" and 1% as "very poor"; a small share of 8.2% regarded their village as "rich" and 7.2% as "quite rich". The relative economic situations of families within a village (A4) was estimated by 65.8% as being "more or less on the same economic level", while about a third (32.7%) perceived a "noticeable difference between the rich and the poor". However, 97.8% of the respondents "mostly perceive[d] equality in the village" (A6) regarding the distribution of power (see figure 5-6 below).[221]

This implies that, even if a third of the respondents perceived a poverty gap in financial terms, they still perceived themselves as being on the same level of influence in terms of power.

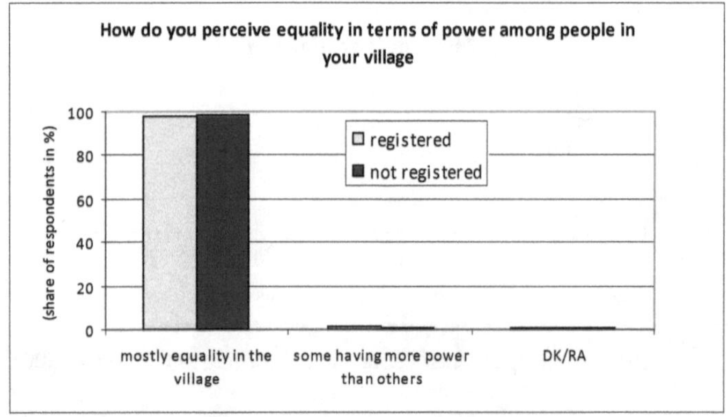

Figure 5-6: Villagers' perceptions of equality within their village, according to CSS survey (2012)

Two-thirds (65.5%) of the population stated that "[t]rust among people is about the same everywhere" (T2), while a third (32%) claimed that "[p]eople trust each other more than in other villages", in contrast to 0.5% of respondents who felt "[l]ess trust in this village". These results are in line with previous accounts of the

[221] The exact wording of question A6 is: "Villages differ from each other in different ways. In some villages there is more equality and cohesion. Whereas in other villages there are groups or people who have more power and more influence on the village life. How would you describe the situation of your village in that regard?" (see appendices).

positive nature of bonding social capital among Georgia's rural dwellers (Buschmann 2008).[222]

Regarding the discussion of common village problems (B1), the majority of respondents usually "gather in small groups" (70.5%) or meet with a "large part of the village (…) to discuss" (24%); only a few (4.8%) expressed that "villagers hardly commonly discuss".[223] About half of the respondents entered regularly into political discussions (R5): 37.5% engaged in political talks "very often", while among 13.5% of the respondents political discussions "always" emerge; in contrast, political discussions are "not so often" participated in by 24.2%, "seldomly" by 15.2% and never for 8.8% (see figure 5-7).

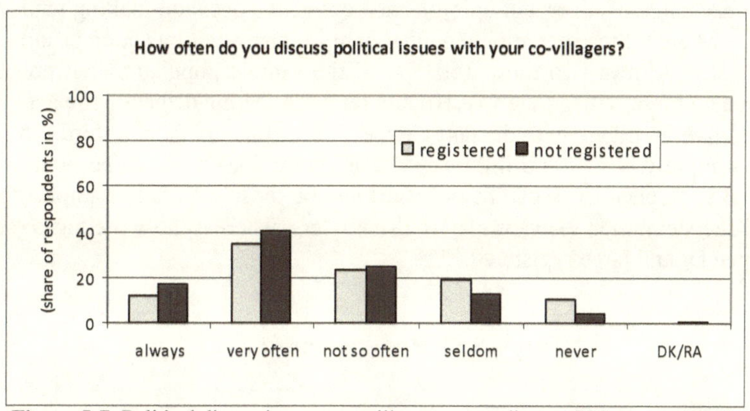

Figure 5-7: Political discussion among villagers, according to CSS survey (2012)

[222] The concept of social capital was first coined by Hanifan (1916) to highlight the productive inclinations of social assets (Putnam 2000: 445). Further elaborated by anthropologist Bourdieu (1980), social capital became defined as a resource shared among individuals and groups to secure a particular social hierarchy and benefit from its corresponding material and symbolic privileges (Méda 2002). Based on reciprocity, norms and specific forms of trust, it also functions as a means for reducing transaction costs (Ben-Porath (1980); Granovetter (1985); Coleman 1988). Meanwhile, Gittell & Vidal (1998) have further developed the idea via their distinction between *bonding* and *bridging* social capital (Szreter 2002: 575–576): Bonding social capital is a feature usually shared in a rather homogeneous social structure characterized by a relatively high degree of "embeddedness" (Granovetter 1985) and "closure" (Coleman 1988) which enables the enforcement of collective norms (Ben-Porath 1980). Bridging refers to situations where individuals stemming from rather different environments need to come together; the "reason for their interaction is to engage together in a collective activity, which each values and benefits from (...), and which is not available through the bonded networks they have" (Szreter 2002: 576). In other words, as Putnam states (2000: 23): "Bonding social capital constitutes a kind of sociological superglue, whereas bridging social capital provides a sociological WD-40".

[223] In cases of damages occurring to some part of their village, two-third of the population (66%) expected help from the majority of fellow villagers, while a third (32.3%) expected help from "[o]nly one part of the village population; many will try to free ride" (A12).

However, what is remarkable is that the discussion of political issues proved to be significant for registration rates:[224] The more the villagers were involved in political discussions, the less they tended to register.

With respect to finding compromises in village situations where "there is divergence of opinions with regard to a problem related to the common interest" (A8), more than half (55.7%) of the respondents stated that "[v]illagers manage to combine different opinions and act unanimously"; by contrast, 30.5% asserted that "[m]any villagers try to promote their opinion and cannot act unanimously", while 11.7% affirmed that "[v]illagers try to let the people with more authority or government representatives take the decision". The way villagers deal with conflicting opinions seem to be reflected by the registration rate:[225] People who observed exchange of divergent opinions and collective decision-making tended to register, whereas those perceiving neither compromise nor joint action usually did not register. Almost two-thirds (60.8%) of the sample population felt pressure from the common village view (A10) and reported feeling a "heavy impact - difficult to live in village with deviant choice", in contrast to about a third (36.5%) of the sample who felt "no influence – everybody does how she/he likes" (see figure 5-8). A clear trend can be observed among those who felt an impact from the village's common view, as they registered less, whereas those feeling no such influence by and large registered.[226]

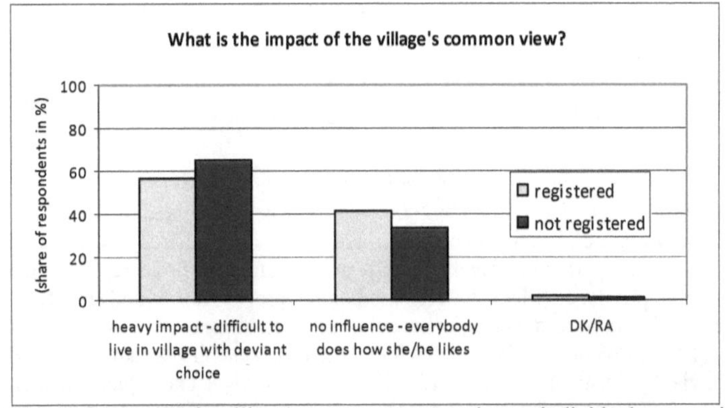

Figure 5-8: Impact of a village's common perspective on individual perspective, according to CSS survey (2012)

[224] The t-test used for this interval scale revealed a probability of p=0.000 and, thus, the relationship proves to be highly significant.

[225] The chi-square test for this nominal scale revealed a probability of p=0.002, indicating significance.

[226] The chi-square test for this nominal scale revealed a probability of p=0.056 and, though it does not indicate strict significance, it supports a clear trend of interdependency between peer pressure and rate of registration.

The country's changes over the past twenty years in comparison to changes in the respondent's village (A2) was viewed ambiguously. It was mostly positively perceived by 42.5% of the respondents at both country and village levels, whereas an almost equal number (42.8%) regarded the country's changes as positive but did not feel any impact at the village level. However, when asked specifically whether urban lifestyle characteristics to are becoming more and more available for village residents within the last five years (A3), 42.3% stated that "[l]ife in the village is becoming more modern", whereas 52.5% felt that "[n]othing has changed in the village in that regard".[227] What is remarkable is that those people who perceived less change in their village registered their land, whereas those who saw life "becoming more modern" did not.[228]

The main source of information (M1-4) for the vast majority of the households was said to be television (92.2%), whereas the press was only mentioned by 8.5% and radio by 3.2%. Stated the other way around, the press and radio are not used at all by most rural dwellers as sources of information (74.3% and 82.8%, respectively). The internet as a source of information is used often by 3.7% of the respondents and seldomly by 7.2%; meanwhile, 89.2% of the rural dwellers did not use the internet at all as a source of information.

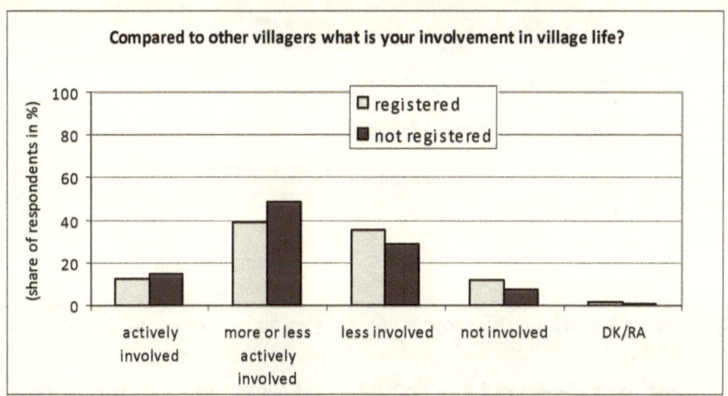

Figure 5-9: Involvement in village life, according to CSS survey (2012)

Collective action in Georgian villages surveyed (R1) could be described as rather vital, as more than half of the respondents viewed themselves being "actively" (13.3%) or "more or less actively" (44.2%) "involved in the common life of the village", in contrast to a about a third who are "less" (31.7%) or "not involved"

[227] Question A3 reads: "As people say, in the modern epoch the services, new types of activities and contemporary life style characteristic to a city are becoming more and more available for village residents. Out of the statements listed below which describes the changes of that kind that have been taking place in your village for the last five years?"

[228] The chi-square test performed for this nominal scale revealed a probability of $p=0.010$, indicating significance.

(9.8%).[229] The relationship between individual involvement in village life and rate of registration proved to be significant (see figure 5-9):[230] Those (more or less) actively involved in village life tend to register less, whereas the less people are involved, the more they register. Also significant is the relationship between registration and respondents who, when somebody sells his or her house/land, preferred the informal rule that they "should offer it to neighbors in the first instance" (B9).[231] In this vein, those who relied less on their neighbors tended to register their land. Here, 57.2% of the sample population recognized this informal rule, while 32% did not, and 10.8% did not know or refused to answer. The majority (70.3%) of the respondents acknowledged another unwritten rule "that almost everyone follows" of "[s]hepherding the cattle one by one" (B10.1), while 20.3% reported that "[t]his problem does not concern our village" and 8.8% asserted that "[t]here are no unwritten rules and customs" (see figure 5-10). Collaboration in the field of herding cattle was, hence, acknowledged by most villagers, though there does not seem to be any co-relation to people's registration behavior.

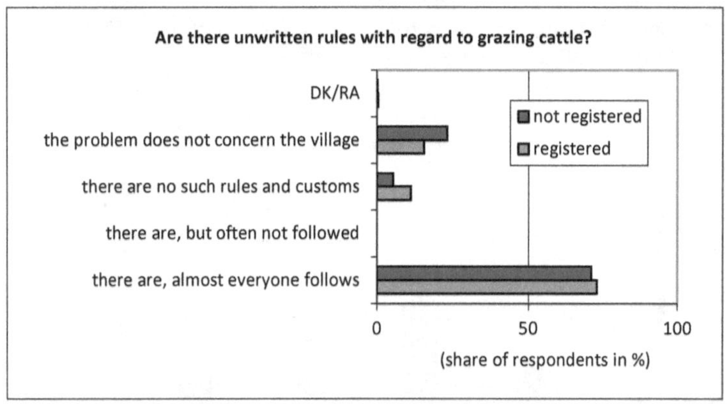

Figure 5-10: Unwritten rules toward grazing cattle, according to CSS survey (2012)

Less than half of the respondents (41.8%) acknowledged an unwritten rule "that almost everyone follows" regarding "using the village pasture (number of cows per village, paying the fee)" (B10.2) and, hence, limiting consumption. Meanwhile, 21.5% knew of "no unwritten rules and customs", while 32.5% asserted that "this problem does not concern our village". It therefore seems reasonable to assume that although shepherding cattle is communally managed among more

[229] For example, 90.8% of respondents (N=600) confirmed participation in celebration of traditional holidays by a large part of the village community (B8).

[230] The chi-square test for this nominal scale revealed a probability of p=0.041, indicating significance.

[231] The original question in full: "Is there the following 'unwritten rule' in your village? If someone sells his/her house or land, he/she should offer it to the neighbors in the first instance." (B9; CSS 2012).

than two-thirds of the rural population on some form of pasture, husbandry on village pastures is often not feasible throughout Georgia, as the population suffers from a lack of pasture land, particularly village pastures.

As far as the enforcement of unwritten rules (B12) is concerned, the majority (73%) of respondents declared that "[t]he village does not have a clear attitude toward such people [i.e. rulebreakers], this kind of behavior is not paid much attention", while 21.8% asserted that "[t]he village is quite strict toward such people"; 5% did not know or refused to answer. To interpret these results is again problematic, seeing that the reply chosen by the majority of the villagers contains two different aspects, namely not having "a clear attitude toward such people" who do not comply with rules or not paying "much attention" to them. Accordingly, the sanctioning of breaking informal rules on grazing is either not enforced equally among the surveyed villagers or is not widely known among them. However, given that a fifth of the sample population (21.8%) reported being "quite strict toward such people", it can be assumed that, although most villagers behaved in accordance with informal rules, "strict" application of sanctioning mechanisms seems limited.

When asked whether villagers provide help for village problems or rather focus on their own family (A7), about two-third of the respondents (68.2%) perceived common efforts within their community, whereas 29.5% saw a focus on the family among the majority of their fellow rural dwellers. The perceptions among respondents regarding collective action seem to mirror registration rates:[232] People who perceived collective action among villagers did not tend to register, whereas those who saw a focus on the family did register.

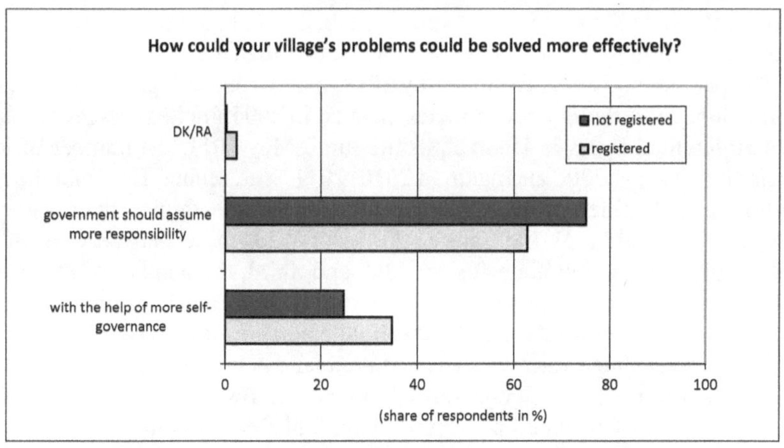

Figure 5-11: Solving village problems, according to CSS survey (2012)

[232] The chi-square test for this nominal scale revealed a probability of p=0.044, indicating significance.

Finally, the majority of respondents (70.5%) believed that a "village cannot resolve village problems, the government should assume more responsibility", while less than a third was convinced that the "residents of the village could manage to resolve [them] if there was more self-governance" (A9; see figure 5-11 above). The relationships between both variables appear to be correlated: those expressing reliance on the government tended to registered their land less often, whereas those believing in the impact of self-governance registered it more.[233]

5.1.4 Summary

The quantitative analysis presented here is based on a CSS survey on "The Role of Social Capital in Rural Community Development in Georgia" (CSS 2012). The study included 600 villagers from 20 villages, located along major roads and in remote areas, chosen from various regions of Georgia by stratified sampling to guarantee inclusion of the so-called village intelligentsia. (The selection process may, thus, have had an influence on the results.) Since questions on (the formalization of) property rights in agricultural land in Georgia were included in the questionnaire, it became possible to understand the characteristics of small-scale farmers, their modes of agricultural production and their impact on land registration. The analysis presented in this section provides general information on property rights formalization (5.1.1), sheds light on the influence of household and farm characteristics (5.1.2) as well as on the impacts of cohesion and collective action in studied villages (5.1.3) on the formalization of land ownership.

As the first sub-section reveals (5.1.1), more than half of the respondents (51.7%) had not registered their land by 2012, mainly due to high costs or lack of information. Even though new policies enacted in 2005 granted lessees and land users preferential rights in land acquisition until May 2011, the number of land registrations rose in 2006 and again in 2010–2011. Yet, seeing, first, that the registration rate identified by the CSS survey is many times higher than was estimated by KfW and the World Bank in 2011 (KfW 2011), second, services of the public registry were only launched in 2008 and, third, the number of those who refused to answer or did not know the time of the registration adds up to 20%, the accuracy of the present rate is to be taken with some caution. Overall, however, the figures reveal huge variations of registered land between and within regions.

According to the latest figures on individual land ownership in Georgia from 2003, the average household possessed about 0.8ha of agricultural land, as presented in 5.1.2. Seeing that this number did not noticeably change for a decade,

[233] The chi-square test for this nominal scale revealed a probability of $p=0.002$, indicating significance.

land tenure for the rural population of Georgia seems to have been constant throughout the years. Based on these figures, the majority lives in a household which is shared by four members where at least some, if not all, of its members are engaged in everyday agricultural work. A household in the West of Georgia is typically located on a larger household plot and is surrounded by orchards and other agricultural products; the possession and working of additional agricultural land plots here is the exception rather than the rule. Meanwhile, a household located in the East is situated on a smaller household plot and usually works several agricultural land plots in addition. (The reason for these two distinct land tenure types goes back to different rural settlement structures, which are subject to analysis in 5.3). Those agricultural areas that stand at a family's disposal are usually completely put to use. The leasing of agricultural land seems to rarely takes place, seeing that 95% of survey respondents stated that the "[f]amily does not lease any land". People prefer to work individually on less land than sharing land and agricultural work with other families; thus, we can say that agricultural work is fully internalized within the family. These findings indicate that agricultural land (including village pastures) is a scarce factor, which confirms similar views from the literature (see chapter 1.2). However, the share of families who do not process any land at all is relatively small, with 1.7% among villagers located in the West and 1.1% in the East. (Whether land resources are more scarce in the West than in East, or better opportunities for off-farm employment exist in the West of Georgia is analyzed in 4.1.2).

Almost one third of the sample population (29.5%) of the CSS survey owned one cow, and a cumulative share of two-thirds of households (69%) held up to ten cows, with a small part (5.7%) of respondents keeping up to 40 cattle; a share of about 25% of villagers did not keep any cattle at all. In monetary terms, almost one third of respondents had an income of 101–200 GEL (29.8%), another third earned 201–300 GEL (31.7%) and 16.2% received 301–500 GEL per month; only 3.7% earned more than 501 GEL. These income levels allowed 54% of the villagers to purchase "food and items of primary consumption", and 26.8% were able to afford "clothes and minor expenses", while 16.7% did not have enough food at their disposal. In contrast, a share of 1.3% stated that "[f]rom time to time we manage to make big purchases", and for 0.3% of the villagers their "[i]ncome is enough for everything", including savings. It is not surprising that income levels significantly correlate with rates of registration: The more people earned, the more probable it was that they registered their land. The survey results also reveal that the nationality and religion of respondents proved to be correlated with property rights formalization: Respondents originating from Armenia and Azerbaijan tended more to register land than their Georgian counterparts; in terms of religion, Orthodox Christians registered less often than Muslims.

Section 5.1.3 reveals a clear correlation between respondents' attitudes toward social cohesion and collective action in their village and formalization of land ownership. First, in terms of describing the overall context, the economic situation for respondents in relationship to others in their village was generally perceived as being similar among the majority; but just under a third perceived themselves as being more deprived than others, whereas about 15% regarded their situation as being better-off. However, almost all of the respondents (97.8%) generally felt there was equality within their village in terms of distribution of power. Thus, it is surprising that, even if almost a third of the respondents did recognize a poverty gap, families perceived themselves to be on the same level of influence in terms of power. In accord with the literature on the bonding nature of the villagers' social capital (Buschmann 2008), trust was equally perceived among the villagers, though a third seemed to feel comparatively more trust in their village than elsewhere. Thus, in situations where opinions diverge among villagers regarding how to solve a village problem, more than half of the respondents were certain that villagers would be able to find a compromise and act together; less than a third was convinced to contrary that no joint action was possible. It was said that common village problems are typically discussed by gathering in small groups, and at least half of the rural dwellers regularly entered into political discussion. Here, it is remarkable that those who appeared to be more involved in political discussion registered less. This could be related to the finding that almost two-thirds (60.8%) of respondents felt influenced by the views commonly prevalent in their village, with a bit more than a third feeling no impact. This paralleled a trend of those claiming not to be influenced by other villagers also tending to have registered their land. More than half of the respondents views themselves as being (quite) actively involved in village life. According to previous findings, more involvement in village life correlates negatively with registering land: the more people are engaged, the less they are likely to register. Further, the majority of respondents were convinced that mutual help among villagers was prominent in their village – and, hence, was correlated with lower registration – whereas a third perceived their fellow villagers being more focused on their family, which was correlated with formalization of land ownership through registration. Nevertheless, the great majority of villagers believed that the government should solve village problems, and only a third was showed a preference for self-governance. Those believing in the impact of self-governance tended to register their land, whereas those relying on the government were less likely to register.

The changes undergone by country were perceived positively by the vast majority of respondents, despite the fact that more than half of them felt no change at the village level. Here it is remarkable that those who did not perceive change in their village did register, whereas those who observed modern urban tendencies at the village level registered less. Television was the main source of information for almost all of the respondents (92.2%), whereas only 3.7% of households used the internet. Seeing that sales of state-owned agricultural land are only announced on the website of the Ministry of Economy and Sustainable Development, it can be surmised that such land-acquisition opportunities are not directed toward Georgian rural dwellers. Meanwhile, although the bulk of respondents recognized informal rules such as informing neighbors first when selling a house or land or shepherding cattle, rules regarding the use of pastures were not approved of by most respondents. Although jointly managed shepherding of cattle was supported by respondents, the data shows that husbandry is not feasible throughout Georgia, due to a lack of pasture land. According to respondents, conforming with informal rules is generally not paid much attention to, and application of sanctioning mechanisms rarely takes place. Nevertheless, according to the survey data, where such rules do exist, only a few do not comply with them, even though "strict" application of sanctioning mechanisms is generally limited.

To summarize, those who have registered land generally have higher incomes and are tend not to be native Georgians (but rather Muslims). They are less limited financially, believe in self-governance, promote unpaid work, acknowledge rules to collect money for the needy and are prepared to work jointly to succeed in agriculture. However, they are also generally less involved in village life – and hence perceive less change taking place in their village – and rather focus on taking care of their family. Those who have not registered tend to be native Georgians (and more likely Christian Orthodox), with lower income levels and convinced that the government should assume greater responsibility to solve village problems. They are actively involved in village life, perceive collective action to be (rather) vital in the village, are more involved in political discussions, feel pressure from the commonly held views of their village and are more sensitive vis-à-vis changes in the village. As a result, it seems plausible to assume that this data on the formalization of land ownership implicitly shows that Georgian villages nowadays consist of two quite-distinct groups of rural dwellers: On the one hand, there is a rather larger group of established villagers who are financially limited from registering land but who are involved traditionally in village life and collective action; on the other hand is a smaller group that is financially endowed and eager to promote alternative ideas to overcome common village problems and succeed in agriculture.

The analysis of the CSS survey presented here suggests that financial means are decisive for those choosing to register property. It also shows that villagers prefer agricultural production internalized within the family rather than potentially achieving higher incomes in agriculture by cooperating with others, and being involved in the village community seems to substitute for seeking sanctuary via so-called secured land titles. On the other hand, about a third of the rural population appears to be open toward new developments and alternative ideas. Whether formalized land ownership provides more (perceived) security for property rights to land and, thus, might lead to increases in agricultural production, is the subject of the next section.

5.2 Qualitative empirical findings

The study commences with a formal analysis of laws and regulations dealing with the acquisition of ownership to understand the evolution of its formal rules and regulations as an outcome of the collective-choice arena. In a second step, its impact on the operational level is then tested on the basis of Grounded Theory (see chapter 2.1.2) and the use of the software atlas.ti (chapter 2.1.3) to finally (5.2.1) understand the evolution of the legal framework to land acquisition, (5.2.2) to analyze the relation between formal (collective choice) rules and local (operational) working rules, and (5.2.3) to evaluate the outcome of the reforms by its impacts on resource users and local production on the ground.

5.2.1 Evolution of the legal framework for land privatization

During the Georgian land reform process (as outlined in chapter 4), a number of laws and regulations were enacted with regard to land privatization (see table 5-12). As most studies have already examined the stages of land distribution and privatization in Georgia up to 2012 (see chapter 1.2), the present study seeks to shed light on the final step of the process: formalization of land ownership. Before turning to the local level to investigate the outcomes of the Government's undertakings, I first outline the circumstances before the new laws were adopted as well as the purposes of the reform process.

As was illustrated in chapter 4 the Georgian government paved the way for property registration beginning in 1996 by introducing a legal framework according to which agricultural land was to be registered in a Public Register (Law of Georgia On Agricultural Land Ownership [1996], Law on Land [Immovable Property] Registration [1996]). The approval and operation of the Civil Code of Georgia of 1997 introduced mandatory property registration. Meanwhile, about a third of new land owners had not yet obtained any documentation to prove their ownership, and transactions involving those who had obtained a document had not been filed. At the end of the 1990s, USAID began systematic registration in parts of Georgia with a limited number of initial free-of-charge registrations and the issuance of ownership certificates (Urgent Measures for the Initial Registration of Agricultural Land Ownership [1999]). The "quick and dirty" approach to surveying and registering land – issuing as many certificates as possible within a relatively short time – lacked appropriate footing (technically as well as organizationally), as it produced erroneous data with drastic implications for registering property and setting up a cadaster in the future.

In the following period, the registration process was simplified and costs reduced. At the time, several state bodies were in charge of land management and land administration, which led to overlapping of competencies, and various cases revealed violation of people's right to register their property. Initial endeavors by multiple donors dealing with land surveyance in Georgia were marked by competition, but later efforts led to the harmonization of data and establishment of the

National Agency for Property Registrations (NAPR). The latter was trusted with property registration beginning in 2004, and two years later it launched the cadaster-creation process. In 2005, the government began focusing on acquisition of leased as well as unused land (Law on State-Owned Agricultural Land Privatization). However, acquiring land ownership from the state was marked by fraud, according to reports by former lessees. As local commissions were founded to decide on recognition of land possessed legally or illegally, landowners were left dependent on their verdicts.

By 2007, the government began concentrating on the formalization of land rights to lawfully as well as illegally possessed land (possible until 2012) and changed the formerly mandatory registration process toward sporadic registration (Law of Georgia on Recognising Property Rights Under the Possession [Ownership] of Physical and Private Legal Entities [2007]). The costs of registering land at the time were estimated to be relatively high for an average rural dweller. Although the Law of Georgia on Local Self-Governance authorized municipalities to manage and lease out pasture land beginning in 2005, five years later the government enacted the Law of Georgia on State Property (2010), which not only set the legal ground to retransfer ownership rights of pasture land back to the national level, but – by having a retroactive effect – allowed acquisition of pastures leased before 2005. (Leasing data for these years provides evidence that such land was by and large held for economic or speculative purposes.) Although legal amendments in 2012 granted primary registration of land free of charge, registration numbers remained low.

The most recent land laws in Georgia were intended, on the one hand, to provide former land users who had not yet obtained any documentation with proof of ownership. In addition, a semi-independent state organization was created to maintain a public registry and cadastre, while procedures to register landed property were simplified and costs reduced. On the other hand, the government envisioned privatizing the remaining agricultural land under state ownership to stimulate the land market and, thus, spur agricultural production. What were the actual effects of the new laws? The following sections seek to shed light on this question.

5.2.2 Analyzing the impacts of land reforms on resource users and agricultural production with the Institutions of Sustainability (IoS) Framework

Based on previous work investigating land privatization in Georgia (see chapter 1.2), the present study has been guided by the following *working hypothesis* (see chapter 1.4): Political reforms tackling land privatization benefited those close to or belonging to informed political circles and who are now better off, measured in terms of high agricultural output and secured allocated property rights. Based on this premise, it would not be expected for "outsiders" – meaning here those not involved in political matters, such as the poor or ethnic and religious minorities – to benefit from the reforms.

To more specifically investigate this hypothesis, the subsequent analysis centers on the following questions:

1. Considering that agricultural production is organized by households within rather homogeneous village structures largely inhabited by Georgian, Armenian or Azeri communities, to investigate the impacts of the reforms on minority versus Georgian-populated rural sites: *Have the latest reforms again favored native Georgians, as the initial land-distribution practices did?*
2. As Georgia's agrarian production is determined by various climate zones that differ from one region to the other: *Were the land-tenure reforms realized in the same way throughout the country?*
3. Given the two main approaches that guided the reform process: *is it possible to disclose different effects stemming, on the one hand, from the "wholesale" approach towards land privatization initially pushed by the government (land distribution) and the "incremental" market approach led by international donors (property rights formalization)?*

Against this background, this part of the study concentrates empirically on resource users and processors by focusing on sectors that depend on the long-term enforcement of land property rights and represent two of Georgia's leading agrarian export segments: wine grapes in East (Kakheti) and hazelnut production in West (Samegrelo) Georgia. The following sub-sections examine how the properties of involved transactions (the contentious initial assignment of rights to agricultural land) have impacted institutional and organizational arrangements (by means of top-down designed collective-choice rules) within the respective action situations.

5.2.2.1 Characteristics of agricultural production and actors involved

As was illustrated above (see 2.1 and 3.3.1), the "Likhi Mountain Range serves as a geographical barrier, dividing the country into eastern and western halves" (World Bank 2012: 3), resulting in diverse climate zones, soil types and land uses. Hence, first, (a) experts were interviewed to portray respective local features of each region, as detailed below. Subsequently, (b) agricultural production of both regions is described, based on responses elicited from farmers and agricultural processors of hazelnuts and grapes in interviews and focus groups in 2013 and again in follow-up meetings in 2014.

a) Geographic attributes: The experts who contributed towards building a picture of Georgia's geographic traits were chosen in the respective field of study on the basis of snowball sampling and eventually stemmed from two think tanks (ISET and CSS), the Division of Human Geography at Ivane Javakhishvili Tbilisi State University, the Georgian Orthodox Church, the Ministry of Agriculture of Georgia, the United Nation's Food and Agriculture Organization (FAO), the German Gesellschaft für Internationale Zusammenarbeit (GIZ; formerly the Gesellschaft für Technische Zusammenarbeit: GTZ), an employee of a Georgian wine producer, as well as another wine company's lawyer.

The experts' responses were classified according to codes referring to 'characteristics of Western//Mingrelian villages', 'characteristics of Eastern//Kakheti's villages' as well as 'characteristics of Azeri//Kakheti's villages' by using the 'code manager' function of Atlas.ti (see figure 5-12 below). The groups' respective quotations are given out by the 'editor'. The following descriptions are summaries of the experts' contributions.

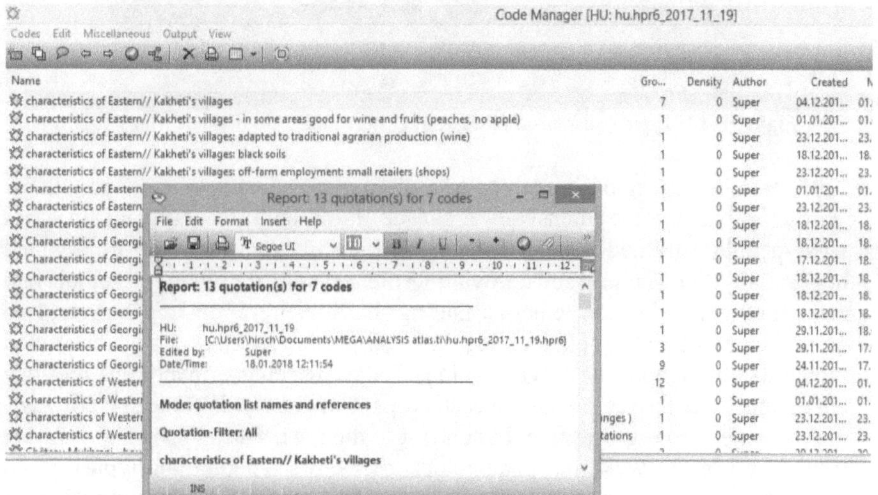

Figure 5-12: Filtering village traits using the 'code manager' function of Atlas.ti

Samegrelo's//Western villages

Due to Western Georgia's rather hilly terrain, agricultural land is relatively scarce, villages are dispersed and, thus, pursuing agriculture is more difficult than in the East. A household plot is rather large and consists of a family's house in a fenced yard with orchards around it (see figure 5-13 below).

Figure 5-13: Typical dispersed village structure, Samegrelo (Google Maps 2017)

Cultivation in this region is adapted to a subtropical climate, where corn and orchards for example are intensively managed in the family's courtyard. In cases where people work additional cropland, such parcels are usually far from where they actually live. Yellow soils provide fertile ground for growing hazelnuts and citrus (i.e. mandarins and lemons). During the Soviet era, this region was used mainly for citrus and tea plantations, the latter producing black tea which gained popularity throughout the Soviet Union. Today, hazelnuts, mandarins and bay leaves constitute the region's main cash crops. Only rarely do some specific varieties of wine grapes grow there. In contrast to the East, sheep husbandry is rarely to be found in the West, but goats, chicken or cows are kept and typically fed around the home, as pastures are rare in this landscape. The villages visited in Samegrelo during my fieldwork contained the following numbers of individuals and households, arable land, perennials and limited amounts of pasture land, as shown in table 5-1.[263]

[West Georgia]	5K	6K	7O
Inhabitants	around 2,000 people, about 500 families	1.500–2000 people, 450–500 families	around 3,000 people, about 1,000 families
Arable land	up to 100 ha	600–650 ha	130 ha
Orchards	1,000 ha	450 ha	1.000 ha
Pasture land	almost none	none	none

Table 5-1: Villages visited in Samegrelo (2013–2014)

These villages contained about 2,000 people, distributed among 500–1,000 families. The majority of land consisted of perennials, and only in a few cases did the amount of arable land exceed that of orchards.

[263] These figures were provided by a member of the local government (*sakrebulo*).

Kakheti's//Eastern villages
In contrast to the West, the topography of Georgia's Eastern region, Kakheti, exhibits rather flat terrain. Villages are much more densely populated and people's parcels are smaller than their Western counterparts. In Kakheti, villages have a linear layout and, thus, houses stand next to the street and their courtyards are situated behind them (see figure 5-14 below).

In contrast to Western Georgia, agricultural land parcels in the east are said to be in vicinity to people's houses. Their household plot is usually smaller but additional crop land and perennials are bigger in size. Areas of cultivation, typically consisting of black, brown or middle-brown soils, are more extensive and perfectly apt to cultivating vineyards or grain in greater quantities Land use in Kakheti varies enormously due to intra-regional climatic differences: For example, in the municipality of Signaghi mainly grains, as corn or wheat, and sunflower are grown. In most areas it is also possible to grow fruits, primarily peaches.

Figure 5-14: Typical linear village structure, Kakheti (Google Maps 2017)

As the area is home to a huge ethnic Azeri population (the region is close to the borderland to Azerbaijan), Kakheti enjoys traditional sheep herding. Azeri communities are said to be market-oriented people. Their communities' social structure, that was recognized also by the Soviet administration, is headed by a so-called *akhzakal*, a kind of respected local leadership, independent of personal wealth, neither elected, nor appointed. In recent times, *akhzakals* tried to protest against the privatization of their villages' pasture land but were threatened by the government and eventually the police. As minorities often face discrimination in Georgia, an expert of CSS states that "they were afraid and just stopped".[264]

264 Interview with Marine Muskhelishvili, CSS on July 12, 2011 in Tbilisi.

The villages visited in Kakheti in 2013 and 2014 inhabit 5,000 to 10,000 people (see table 5-2).[265] The area is marked by arable land which is most notably situated nearby, only a minor part is located within the villages. Pasture land is almost absent.

[East Georgia]	1Q	2K	3S	4G
Inhabitants	8,500 people, 2,000 families	11,801 people, 3,300 families	5,000 people, 3,000 families	6,000 people, 900 families
Arable land	300 ha within the village, plus 2,000 ha nearby	2,300 ha	500 ha within the village, plus 250–300 ha nearby	400 ha within the village, plus 2,000 ha nearby
Pasture land	80 ha (20–40 people bought 300–400 ha)	none	none	100 ha

Table 5-2: Villages visited in Kakheti (2013–2014)

b) Traits of agricultural production: Respondents who were interviewed and participated in focus groups in 2013 and 2014 were grouped into [east] or [west], according to their geographic location – meaning whether they stemmed from Kakheti, Eastern Georgia or Samegrelo in the West – by using the family code manager function of Atlas.ti (see figure 5-15).

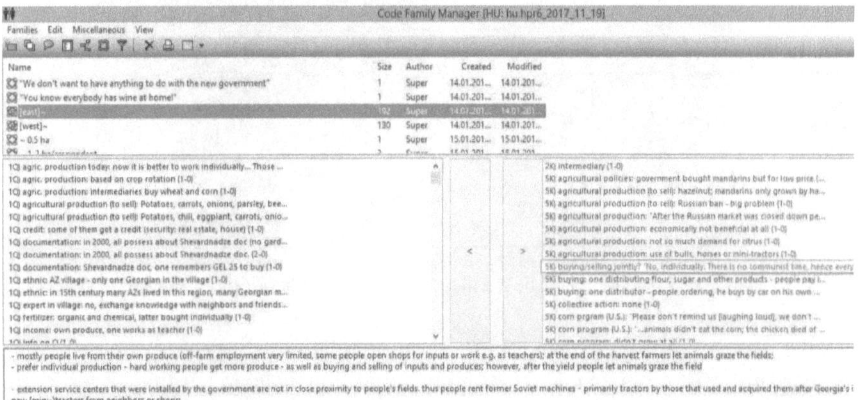

Figure 5-15: Atlas.ti's familiy manager: [east] vs. [west]

General agricultural production patterns

During the Soviet era, people worked their 0.25 ha sized household plot plus additional land of one or at most two hectares on average (mainly stretched over one plot); this land was then distributed or taken over (formally or informally) in the aftermath of the Republic's independence. The majority of people have generally lived from their own produce, for off-farm employment is very limited: some are

[265] The numbers were provided by one member of the local government, *sakrebulo*.

employed on vineyards during the grape harvest, some run routes in areas that were previously unserviced by public transportation by mini-bus (*marshrutka*), whereas others own a shop for inputs or work other jobs, e.g. as teachers. As described in 6.1.2, rural inhabitants in Georgia prefer individual production – a perspective illustrated by one respondent asserting that "hard working people get more produce" or another stating that "[t]here is no [more] communist time, hence everyone makes it on their own". However, due to lack of capital and machinery, they exhibit a noticeable degree of responsibility for each other.[266] Extension service centers, which have only recently been installed by the government, are in almost all cases not in close proximity to people's fields; thus, they either rent former Soviet tractors from those who operated, maintained and acquired them after Georgia's independence or rent (mini-)tractors from neighbors or shops (against payment). Smaller plots of land – e.g. one village possesses about 70 ha agricultural land shared among 150 families – are simply worked by using bulls and horses. Also, people prefer buying inputs and selling produce on their own. If not produced on their own, villagers buy inputs locally from neighbors or shops; larger purchases are made for individual purposes only, such as buying fertilizer for half a year's consumption. Although former agronomists are rarely on-site, knowledge about best agricultural practices is shared among neighbors, with information being spread throughout the immediate vicinity at local central points or ceremonies, such as funerals or marriages. Whereas most groups emphasize having a good relationship to local authorities, they generally reject any contact with the central government.

Agricultural production in Samegrelo

Due to its favorable climate, mandarins, bay leave, kiwis and fennel belong to the region's main produce after hazelnuts. During the Soviet era, kolkhozes and later sovkhozes produced tobacco and silk as well as oils generated from roses and geraniums, and hazelnuts were only grown privately. However, due to climatic conditions as well as villagers' preferences, only half of them plant citrus. Another reason for the reduction of mandarin production was the Russian ban on Georgian agricultural produce (2006–2013), which was said to be such a large problem that people cut down their citrus trees and, instead, planted hazelnut or laurel trees. In fact, it was said that up to three-fourths of the villagers converted their crop land into orchards, as planting corn became less profitable. In some cases, common pasture land was sold and, thus, people lease pastures from individuals. Hazelnuts are delivered by individual growers to processors or sales points; only case reported of an intermediary who comes and picks up the produce from people's homes. In contrast, it was learned that USAID's hazelnut program (2011–2014)

[266] Neighborhoods provide their inhabitants with information, assistance during harvest times as well as rendering (financial) help whenever needed (Buschmann 2008: 12).

provided logistical assistance to farmers in cooperation with the Ferrero corporation. Those included in the program benefitted from seasonal training sessions and establishing direct links to hazelnut processors. Meanwhile, mandarin growers received assistance from the government which granted fixed prices and bought produce for a low price – according to respondents GEL 0.45/kg [USD 0.18/kg] – even though quality did not play any role, seeing that rotten fruit was also sold to the government.

Agricultural production in Kakheti

Whereas irrigation was provided throughout the region during the Soviet era, today water is only available to those having either groundwater pumps, fields nearby a river or an old Soviet irrigation canal and who can afford to pay for access to water. During my field research, I found out that one of the two ethnic Azeri villages had 90% water coverage by sharing drip-irrigation pumps among two to three households, while another did not have access to water for agriculture at all. Among the two Georgian villages, the picture looked the same, with one village enjoying individual water supply via pumps and the other left without water, due to missing electricity infrastructure in this area. Under the Soviet regime, although people did benefit from inputs from kolkhozes and sovkhozes, they could work their own land, the so-called household plot, only at night. Moreover, in those times, machines could not be leased; hence some state that the situation is better today than during the Soviet era. These days, people sell small quantities of their agrarian production at local markets, while larger amounts are brought to the central market in Tbilisi, where greater storage facilities exist. A limited amount of fruits and vegetables that do not normally result in high profit margins, such as garlic, beetroot or potatoes, is grown for the family's own consumption. In contrast to former times, respondents report that more and a greater variety of crops are produced today. The main cash crops, wheat and corn, are sold by individual families to intermediaries who are from the same region and have been known for years. In addition, intermediaries are interested in livestock. Although there is not much communal pasture land left to let animals graze, families are engaged in animal husbandry, meaning chicken, cows, sheep and in some case buffalos, and most own one cow. Animals are usually kept on the home plot or village pastures and, at the end of the harvest, graze the surrounding fields.

Grape distribution to and cooperation with wineries is organized in the following way: A winery employee comes to a family's home and they agree on quantities to be delivered; after the grape harvest each household delivers their own produce individually to the wineries. While some respondents focused on how the seven-year Russian ban – a consequence of widespread impurities in and falsification of Georgian wine and mineral water – led to many Georgian companies going bankrupt and then having to sell their land, others emphasized its positive effects, especially the subsequently tight control of wine quality by Samtrest, a newly found governmental body for monitoring grape quantity and quality just after gathering (and, hence, collecting sectoral aggregate data).

5.2.2.2 Properties of actor transactions

As my main research question is centered on the outcomes of reforms targeting privatization of agricultural land in Georgia, the next task here is "to identify and order heterogeneous transactions" (Hagedorn 2008: 374) in domains where transfers of land resources from state to private ownership have taken place. According to Hagedorn, the three critical dimensions indicating the properties of transactions are (i) *uncertainty*, (ii) *frequency* and (iii) *asset-specificity* (see chapter 3.4.2). The key attributes of ordering and establishing causal relationships between the properties of a transaction in natural systems are *structural modularity* versus *decomposability*, on the one hand, and the *functional interdependence* of processes on the other (see chapter 3.4.2). Hence, the present section investigates the nature of transactions performed (i.e. their underlying organizational form, namely whether comprised of atomistic or interrelated components) as well as the nature of their functional interdependency toward other (sub-)components in the given social-ecological system (isolated versus interconnected), here the rural regions of the Republic of Georgia. For this purpose, the following questions need to be addressed (Hagedorn 2008: 364): (a) The primary question relates to the physical entities affected by a transaction and the basic attributes assigned to it; (b) the next question involves identifying the properties of the transaction that stem from these entities' attributes; and (c) the final question is concerned with how the properties of transactions impact the institutional and organizational arrangements that govern them. These concerns are explored in more detail in the following.

(a) Land is the only commodity comprised of two particular *physical features*: it is immovable and everlasting (see chapter 3.1.4). As a *social artifact*, land as property is grounded in both its function as an *asset*, based on its character as a renewable resource stock, as well as a *flow* in the form of its utility (*Nutzwert*), that can either be exploited directly in the form of yields or indirectly in the form of land rent (see chapter 3.1.3). In economic terms, land is unique for it constitutes a production factor which cannot be multiplied and forms the least expensive way for agrarian production and other human needs (Kuhnen 1982: 70).

(b) Considering the high degree of scarcity of usable land resources in Georgia (see chapter 1.2), transactions related to the (initial) assignment of private property rights to agricultural land are presumable highly contentious in nature, especially as the Republic is characterized by various climate zones and favorable soil types which enable a variety of agrarian land uses. (i) Given these conditions, particularly during the transition people were exposed to a rather high degree of *uncertainty* (see chapter 1.3), specifically as the process of initially assigning private property rights to agricultural land in the beginning of the 1990s was accompanied by civil unrest and political upheavals (see chapter 4.1). Also lack of credible commitment on the side of the government, characterized by steady, deliberate production of uncertainty as the very means to secure power (see chapter 1.2), opened a gateway for opportunistic behavior (see chapter 4.2). This created an agency problem (ex-post contractual moral hazards) where informed individuals – by enacting ambiguous legal acts and implementing policies with a lack of precision (see below) – were able to influence the distribution process of land at the expense of the vast majority of villagers, who should have more land at their disposal today (see chapter 5.1.1). Assuming that high perception of insecurity and risk biases people's conduct (by increasing the discount rate) toward short-term investments (see chapter 3.1.3), a low registration rate for property rights to agricultural land might not only be due to high costs relative to villagers' low endowments but a rational form of myopic behavior in response to limited predictability and perceived insecurity of land titles. (ii) *Frequency* is of considerable concern, seeing that a change of property regimes from state to private ownership is (usually) a one-time process, involving a series of non-recurring transactions, especially compiling the register and cadaster; it thus entails considerable set-up costs, both social and economic (see chapter 3.4.2). It seems reasonable to emphasize (iii) *asset specificity* as the most striking feature of the transaction properties related to property in land. In particular, *site-related* aspects, like soil quality or availability of water, play as much a role as location, meaning here accessibility of infrastructure and proximity to markets and towns. In addition to these spatial particularities, *physical* asset specificity stems from *nature*-related attributes which, above all, are dependent on natural processes and hazards, say soil erosion or floods, as well as from social artifacts, time and scale; particularly time lags and economies of scale affect the degree of (future) benefit streams from land use. With respect to Georgia's formalization of land property rights, physical asset specificity looms particularly large due to *technical* asset specificity, particularly in terms of the set of complex interdependent transactions for compiling and maintaining a modern land administration system (see chapter 3.2.3). Meanwhile, *human* asset specificity is pronounced, given that the concept of private property based on formal contractual relationships and the involvement of state authorities is a rather novel social form within the history of the Georgian Republic, where informal contractual relationships – characterized by oral tradition and reputation – have always played a key role in political circles (see chapter 1.2) as well as among villagers (see chapter 5.1.3). Transacting property rights is also a process highly characterized by *jointness*, for a change in the entitlement structure alters the dual nature of the legal correlates underlying any property relation: right–duty, privilege–no-right, power–liability, immunity–disability (see chapter 3.1.4). As legal procedures related to rights to land are entwined with fiscal and regulatory means within the general framework of public policy, specifically land administration (see chapter 3.2.3), the transactions involved indirectly depend (and have effects) on their degree of *coherence* with other public policy measures, including taxation or protection of minority rights. Overall, the transaction properties are strongly characterized by *complexity*, seeing that the (initial) assignment of rights to land is linked to the emergence of transaction costs (here to be viewed in the context of transition, friction and coherence costs) and, consequently, is a process that impacts the distribution of wealth among the members of society.

c) Legal requirements and procedures vis-à-vis the transfer of rights to land – marked by a high degree of uncertainty, social and economic (set up) costs as well as multifaceted asset specificity – were altered over time (see chapter 4.5) in Georgia and, with the more recent introduction of a modern land administration system (see chapter 3.2.3), have increased in complexity. Based on the outcomes of formerly enacted policies targeting land privatization, rules were modified and changed over time. At first, legal acts targeted transfer of ownership rights to a limited quantity of "reserve land", reportedly marked however by incomplete conveyancing of people's rights and titles. A few years later came the legal assignment of use rights to agricultural land and, after ten years passed, new rules initiated the private acquisition of such land by granting primary purchasing rights to former leaseholders (pasture land excluded). At the same time, procedural rules were introduced for marketing the remaining land resources belonging to the state. Steered by the international donor community, the formerly widely distributed and dysfunctional administration of the Republic's diverse land resources was finally linked by updated organizational procedures and integrated into the public registry, leading to a renewal of governance structures for agricultural land (see chapter 4.4). As these events were linked by both rule-changing situations and organizational procedures, I have conceptualized the policy process of privatizing land into a complex chain of four action situations (see chapter 3.4.3), based on their primary type of transaction: (I) *distributed land*, (II) *leased land*, (III) *legally or illegally possessed land,* and (IV) *privatized land* (an overview of the particular laws and regulations targeting land privatization is provided in table 5-3 below):

I. *Distributed land*: Primary transfer of use rights to agricultural land, distributed from the state to mostly rural households: Resolution No. 48 (1992); Resolution No. 128 (1992); Decree No. 503 (1993).

II. *Leased land*: Transfer of use rights for a 49-year term: Law on Agricultural Land Leasing (1996); Presidential Decree No. 446 (1998). From 2005, lessees were granted rights of preemption with a 5-year window up to 2011, together with preferential payment schemes; but even though the acquisition of pastures was legally strictly prohibited by the Law on State-Owned Agricultural Land Privatization (2005), an additional retroactive clause in Article 4.1 (bb.a) of the Law of Georgia on State Property in 2010 provided for the privatization of pasture land leased out before 30 July 2005.

III. *Legally or illegally possessed land*: The Law of Georgia On Agricultural Land Ownership (1996) stipulated obligatory registration of land and became in 1997 constitutionally compulsory, though no organizational structure had yet been launched to enforce the formalization of land rights. Presidential Decree No. 327 provided for a limited number of systematic free-of-charge registrations (1999–2003). In 2007, these rules were modified toward (voluntary) sporadic registration in the Law of Georgia on Recognising Property Rights Under the Possession (Ownership) of Physical and Private Legal Entities and saw last changes in 2008 with the Law of Georgia on Public Registry. In 2011, the government enacted Regulation No. 509, stipulating the per-unit registration price for immovable property at GEL 50. Legislative changes in 2012 set primary registration and registration for specifying the area as free of charge in the Ordinance No. 231 on the Regulation of Certain Issues Related to the Registration of Titles (2012).

IV. *Privatized land*: Since 2005, the Ministry of Economic Development of Georgia (MoE) has been in charge of privatizing state-owned agricultural land via special auctions, open auctions, or formerly leased land through direct sale under the Law on State-Owned Agricultural Land Privatization (2005).

As emphasized by Lasswell (1971: 2) and others, "norms of conduct are partly determined and made effective outside the machinery of legislation, administration and adjudication, which, once again, stresses the need to study the actual rules-in-use" (Theesfeld et al. 2017: 4). As legal rules were set by the Georgian government to impact current and potential resource users, the next section focuses on authoritative relationships (Commons [1924] 1968) that "affect operational action situations, called collective choice rules" (Ostrom 2005: 187; see chapter 3.4). Applying a vertical approach makes it possible to "recognize that rule sets are themselves nested in hierarchical levels. The participants in operational situations are directly affected by the operational rules structuring what they must, must not, or may do" (ibid. 2005: 215). The following section hence aims to identify the effects of collective-choice rules on resource users and their working rules.

Time of enactment	Laws and regulations	Source of legal authority
January 18, 1992	Resolution No. 48	Cabinet of Ministers
February 6, 1992	Resolution No. 128	Cabinet of Ministers
June 28, 1993	Decree No. 503	Cabinet of Ministers
August 24, 1995	Constitution of Georgia	President of Georgia, E. Shevardnadze
June 28, 1996	Law on Agricultural Land Leasing	President of Georgia, E. Shevardnadze
March 22, 1996	Law of Georgia On Agricultural Land Ownership	Parliament of Georgia
1996	Law on Land (Immovable Property) Registration	Parliament of Georgia
June 26, 1997	Civil Code of Georgia	President of Georgia, E. Shevardnadze
May 16, 1999	Presidential Decree No. 327, Urgent Measures for the Initial Registration of Agricultural Land Ownership Rights and Issuance of Registration Certificates to Georgian citizens	President of Georgia, E. Shevardnadze
July 8, 2005	Law on State-Owned Agricultural Land Privatization	President of Georgia, M. Saakashvili
December 16, 2005	Organic Law of Georgia on Local Self-Government	President of Georgia, M. Saakashvili
July 11, 2007	Law of Georgia on Recognising Property Rights Under the Possession (Ownership) of Physical and Private Legal Entities (abolished Decree No. 327)	President of Georgia, M. Saakashvili
September 19, 2008	Law of Georgia on Public Registry	President of Georgia, M. Saakashvili
August 9, 2010	Law of Georgia on State Property	President of Georgia, M. Saakashvili
December 17, 2010	Law of Georgia No 3889	President of Georgia, M. Saakashvili
December 29, 2011	Regulation No. 509	Government of Georgia
March 28, 2012	Law of Georgia No. 5928	Government of Georgia
June 28, 2012	Ordinance No 231 on the Regulation of Certain Issues Related to the Registration of Titles	Government of Georgia
June 3, 2016	Law of Georgia on the Improvement of Cadastral Data and the Procedure for Systematic and Sporadic Registration of Rights to Plots of Land within the Framework of the State Project	President of Georgia, G. Margvelashvili
	= white rows refer to procedural rules targeting land privatization	
	= green rows refer to the registration of land titles	

Table 5-3: Overview of laws and regulations on land privatization and registration of Georgia (1992–2016)

5.2.2.3 Institutions for land-related transactions

Having established the properties of the transactions relevant to land property rights in Georgia, we now need to comprehend the working rules used by individuals involved in those transactions (see chapter 3.4.3). The tasks at hand are therefore (1.) tracing *collective choice rules* associated with land privatization according to the four types of land resources identified above (I–IV); (2.) outlining the *operational rules* underlying the transfer of land resources from state to private ownership; and then (3.) *comparing* the formal rules with those actually used on the ground. The three-part typology developed by Cole (2017) is applied to classify the relationships between formal legal rules and working rules (see chapter 3.4.3), to wit assessing whether the formal rule is similar to the working rule and thus enforced as written, whether the formal rule has an effect on the working rule and/or vice-versa but the degree of enforcement is low, or no apparent relation exists between the two and, hence, no enforcement is to be anticipated; in the latter case, the legal rule might not be more than a symbolic act.

I. Distributed land

1. *Collective-choice rules*: During the Georgian transition, agricultural land was distributed to households out of a privatization fund by local land commissions, according to the following criteria (see chapter 5.2):[267] (i) The amount of land distrubuted accounted to 1.25 ha of land to former kolkhoz or sovkhoz staff residing in the lowlands, 0.75 ha for other rural households and up to 5 ha for families living in the highlands; urban settlers obtained 0.25 ha of agrarian land; (ii) regading the number of distrubuted land plots households received on average three or four different, often dispersed plots, and in some cases land was also distributed to individuals; (iii) as far as documentation is concerned, the majority of new land owners did not receive any proof of title.

2. *Operational rules*: Using Atlas.ti a new family was created called 'I. Distributed land' which is used as a 'global filter'; the code manager then lists all codes associated with land distribution and, by requesting an output, edits the respective quotations (see figure 5-16 below).

(i) The following working rules were applied to allocate land from the state to the people: In general, people kept the same land that they had worked during Soviet times (0.25 ha) and obtained more as well. Accordingly, in Samegrelo for example, people received orchards when they had worked on orchards before, in addition to cropland. Respondents unanimously reported that villagers received 1 ha of land; if the village, however, had more land available, then they received more land, about 1.5 to 2 ha in total, depending on the size of the village. Additionally, during Zviad Gamsakhurdia's reign (May 1991 through January 1992), respondents reported a six-month time span during which they could reclaim land that had belonged to their ancestors. According to one respondent, who still works the land of his ancestors, his "ancestors had vast fruit orchards that were bought with golden money", while he himself now only owns 1 ha (laughs).[268] Meanwhile, both ethnic Azeri villages in Kakheti exhibited smaller landholdings, with most people being said to own 0.5–1 ha consisting of one plot and "only some people have more". Additionally, about half of the respondents there reported their land plots being distant from their village, from 6 up to 20 km. At the time

[267] Resolution No. 48 (1992); Resolution No. 128 (1992); Decree No. 503 (1993).

[268] This exceptional case of restitution in Samegrelo was confirmed in an interview on July 15, 2013, by a land tenure expert, GIZ, Georgia, Tbilisi.

of my fieldwork, all other respondents possessed about 1 ha of agricultural land of very good quality – the only outlier, an expert in animal husbandry, owned 37 ha of pasture near a village where pasture land was scarce.

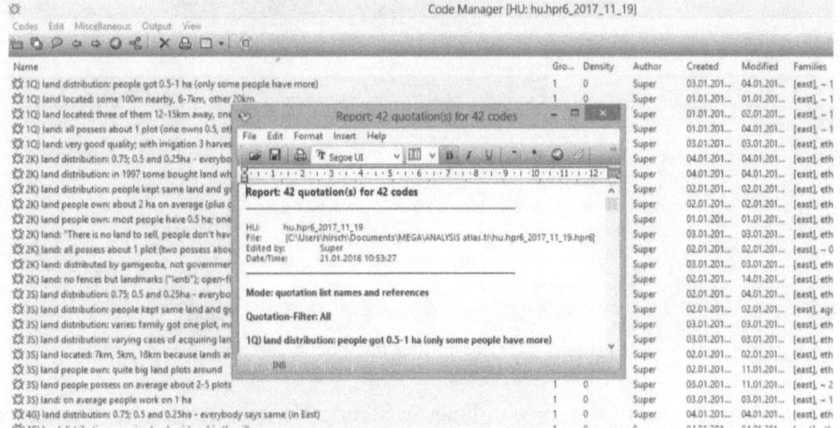

Figure 5-16: Global filter for 'I. Distributed land'

(ii) Tracts of land are not fenced in these regions, but people use landmarks, called *lenti*. Some reported using an open-field system after harvesting. The average landholding of respondents stemming from the ethnic Georgian villages was 1–2 ha per household at the time of the study, with people owning two to five land plots that are distantly located from the village. In one of the two villages, it was said that land in this region was distributed according to the availability of fertile soils in the villages. Families usually received one plot according to respondents in Samegrelo and in Kakheti, whereas land allocated to individuals consisted of scattered plots. The latter, however, seems to have been rare. Even though the cases under investigation exhibit smaller landholdings for ethnic Azeri villagers (0.5–1 ha/household) than of ethnic Georgian villagers in Kakheti (1–2 ha/household) and those in Samegrelo (1.5–2 ha/household), more comprehensive data is needed to support the claim that ethnic Azeris were disadvantaged in the allocation process. The talks revealed a general lack of land in the region, whether pasture or agricultural land.

(iii) The question concerning documentation of land ownership required another query using Atlas.ti, where a new family needed to be created comprised of all codes related to 'documentation'. By using a global filter, the respective quotations were then edited as output via the code manager (see figure 5-17).

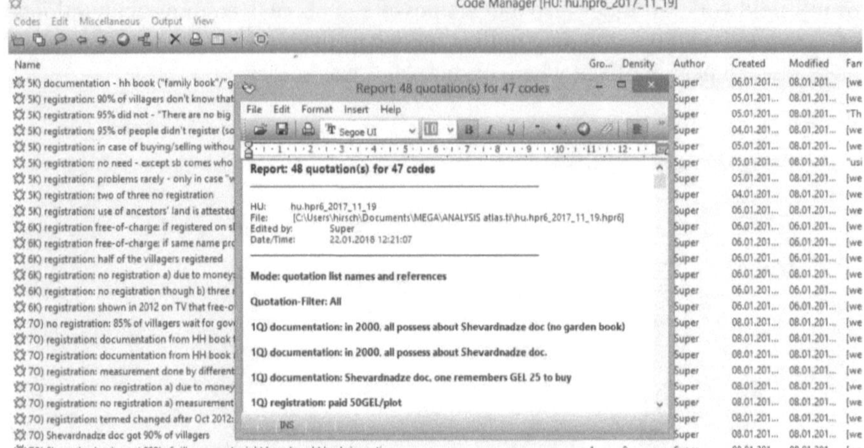

Figure 5-17: Global filter for 'documentation'

Respondents in one of the three villages in Samegrelo declared that they obtained land that was registered in the garden/household book and, hence, possessed documentation stored in the local archive (here, the former regional archive in Kutaisi). Thus, if any documentation for Respondents in one of the three villages in Samegrelo declared that they obtained land that was registered in the garden/household book and, hence, possessed documentation stored in the local archive (here, the former regional archive in Kutaisi). Thus, if any documentation for the land were to be required they would go to the archive and from there take the papers to the Public Registry. Respondents in two of the other two Georgian villages declared that about 90% of the villagers received a Shevardnadze document. Both focus groups in the ethnic Azeri villages in the East mentioned when asked about proof of ownership that they had expected the land to have already been registered and, hence, properly documented, as Germans had made orthophotos of Kakheti's agricultural land with helicopters in the early 2000s. In one village respondents reported that all households received a Shevardnadze document – with one remembering that they had to pay GEL 25 for it – whereas in the other village respondents claimed that all had registered (due to the Germans) but mentioned the need to get proof from the archive – actually a sign that they have no proper documentation in their possession but that it is at least stored in a local archive. The respondents of one of the two ethnic Georgian villages reported that they had obtained the Shevardnadze document free of charge while, in contrast, respondents from the other village remembered that they had to pay money to receive some kind of documentation that nobody, in fact, ever received; hence, they remain without legal proof of title.

3. The foregoing *comparison* suggests that the relationship between formal legal rules for allocating land to households in Georgia and their interpretation

through working rules has been marked by a high degree of incongruity, seeing that enforcement levels vary not only from region to region but also from site to site. Whereas farmers in Samegrelo received one hectare plus more land if available and benefited from a six-month restitution period during the mid-1990s, ethnic Azeris residing in Kakheti received smaller land plots located more distant from their villages than those of ethnic Georgians residing in the same area who, it seems, got one hectare plus more land if locally available. It thus seems that social norms dominating in Georgia, namely preferential treatment of people of the same origin, had a strong influence on the law's implementation when distributing land.

II. Leased land

1. *Collective-choice rules:* In 1996 a market-oriented reform commenced with the transfer of use rights to land that was made available for leasehold for a period of 49 years (see chapter 5.3).[269] From 2005 up to 2011, lessees were granted preemption rights for direct sale, together with preferential payment schemes. Meanwhile, further means for privatizing formerly leased land if leaseholders were not interested in acquiring it were instituted in the form of special auctions, open auctions or direct sale (see section 'IV. Land privatized' below).[270] The outcome of the reform – about 55% of the total arable land and 68% of the total perennials were privatized – was reportedly biased, as the process was said to be corrupted by influential individuals – members of the local elite and former state and collective farms – who kept most of the land to themselves either for generating a steady flow of income through sub-renting or holding for speculative purposes. In particular, it is said that (a) the majority of the (more valuable) land that was privatized or leased lies near settlements in important agro-ecological areas, and (b) less state-owned land, "reserve land" of the privatization fund, has been available for lease. Legal changes in 2005 most probably preserved the status quo where the majority of available and valuable land is held by a few, formerly influential individuals.

2. *Operational rules*: Another family for Atlas.ti called 'II. Leased land' was formed comprised of codes related to leasing land. Using this family as a global code, all relevant quotations were then listed by the editor (see figure 5-18 below).

(a) Respondents in Samegrelo did not indicate that land in their possession lay far away from their villages; in contrast, about half of the respondents in both ethnic Azeri and Georgian inhabited villages in Kakheti reported their land plots being distant from their villages, from 5 up to 20km.

[269] Law on Agricultural Land Leasing (1996); Presidential Decree No. 446 (1998).

[270] Law on State-Owned Agricultural Land Privatization (2005).

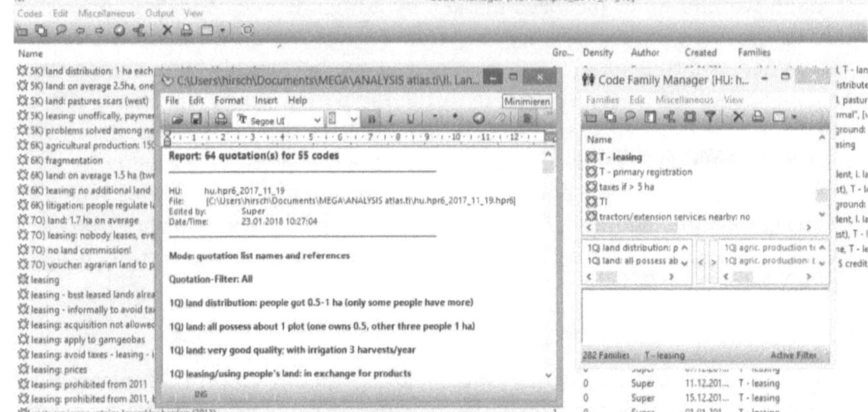

Figure 5-18: Global filter for 'II. Leased land'

(b) The lack of available land resources for lease was of great concern in both regions. Respondents in the three Mingrelian villages unanimously describe a lack of land resources, namely pastures as well as agrarian land; the conditions in one of the villages is exemplifying where 150 families work 70 ha crop land. Moreover, people criticize the lack of state-owned land for lease.[271] However, people of all villages repeatedly indicate that (additionally to land allocated to households in the 1990s) "using others' land is normal". There is hence an indication of certain parcels that are privately owned but left fallow and hence available for use. The situation in Kakheti looks similar but shows different nuances. There are land resources to lease from the state but these are limited and no more land is available. Respondents in one of the ethnic Azeri populated villages state in particular the scarcity of pasture land who thus leases pastures from other individuals – it is the very village where one of the respondents, the outlier, an expert in animal husbandry, bought formerly leased land in 2008 (at that time the acquisition of pasture land was legally prohibited).[272] In both ethnic Georgian dominated villages the picture looks similar, some people lease land from the state or use others' land in exchange for products and machinery. (The surrounding area of the villages is characterized though by huge land parcels that reportedly belong to someone from the government.)

3. The *comparsion* shows that the legal rule to lease agricultural land created a window of opportunity for opportunistic behavior of a few well-informed and influential local individuals. Moreover, the outcome seems to have impacted further juridical developments which in 2005 led to perpetuating the status quo ante by

[271] A representative of the Regional Government (*Sakrebulo*) of Zugdidi stated that "…most of the lands are already sold on the auction, 60–80% in Samegrelo", interview on July 10, 2013, Zugdidi (see section '*IV. Land privatized*' below).

[272] See Law on State-Owned Agricultural Land Privatization (2005).

providing the lessees a right of preemption to acquire the land in leasehold. Rules-in-use to tilling others' fallow land or working others' land in exchange for in-kind contributions help to maneuver around the constant lack of land reserves.

III. Legally or illegally possessed land

1. *Collective-choice rules*: As no organizational structure for land registration had been legally established before 2007 (see chapter 5.4), only a few individuals benefited from limited systematic and free-of-charge registration at the end of the 1990s (see chapter 5.4).[273] (a) Legal changes in 2007 led to the introduction of voluntary sporadic registration at the National Public Registry (NAPR), which was legally founded in 2008.[274] The new law targeted two distinct kinds of land resources, namely those in use before 2007 that were either in *lawful* or *unlawful possession*. This included *type I* land resources listed above that were allocated to rural households in the beginning of the 1990s. *Lawful possession* refers to state-owned land whose proof of ownership has been verified; the plot is registered by presenting proof of title together with a cadastral measurement plan of the parcel and the fee paid at the Public Registry (Art. 2a).[275] *Unlawful possession* refers to land plots and smaller land parcels adjoining a land plot (held under lawful possession) whose steady use has been proven by witness testimonies (Art. 2c). The NAPR is entitled to legalize land held in lawful possession; special (land) commissions, formed by local so-called self-governmental bodies, are entitled to legalize unlawful possession of land. If the commission legalizes the use of land, an ownership certificate is issued that serves as basis for registering the adjacent land plot at NAPR.[276] The recognition of rights to lawfully possessed and squatted land for legal persons was legally prohibited beginning in 2012 (enforced by Law in

[273] Law of Georgia On Agricultural Land Ownership (1996); Presidential Decree No. 327 on Urgent Measures for the Initial Registration of Agricultural Land Ownership Rights and Issuance of Registration Certificates to Georgian citizens (1999).

[274] Law of Georgia on Recognising Property Rights Under the Possession (Ownership) of Physical and Private Legal Entities (2007); Law of Georgia on Public Registry (2008).

[275] Moreover, the Law (2007, Art. 2a) refers to "land squatted before 1994, and registered in a technical inventory archive".

[276] The fee for lawfully possessed land is "transferred to an interested natural person free of charge" (Law of Georgia on Recognising Property Rights Under the Possession (Ownership) of Physical and Private Legal Entities (2007), Art. 6.1) and for legal entities "five times the amount of the annual rate of property tax" (Art. 6.2). The fee for unlawfully possessed ("squatted") land for natural persons is "10 times the amount of the annual rate of land tax" (Art. 6.3b); for legal entities it is "100 times the amount of the annual rate of land tax" (Art. 6.3a). The fee for lawfully possessed land for legal entities changed from 2007 to 2011: "From 1 July 2011, the fee payable for the recognition of the property right to land lawfully possessed (used) by a legal entity under private law shall be equated to the fee payable for recognition of the property right to land squatted by a legal entity under private law" (Art. 7[3]).

December 2010).[277] (b) Legal changes in 2012 led to primary registration and registration for specifying an area free of charge.[278] (c) In 2013, the parliament set up a sectoral voucher program for so-called minor landowners who had registered a maximum of 1.25 ha land, for rendering assistance for ploughing and harrowing of their land or to buying fertilizers and other inputs. The outcomes of this program thus provide an indication of the role land registration plays today.

The following outcomes are known from the literature: The majority of agricultural land distributed in the 1990s is not registered. Only about 30% of the land is registered and, thus, land not registered might be registered in someone else's name, such as by the state as the new owner. Earlier property-right violations were geographically concentrated and related to the local administration of towns or regions, when for example in 2007 private owners apparently gave their property as a gift to the state or in 2009 when the state registered land on its name in potentially touristic zones and subsequently sold it to a touristic company, even though the land had already been registered by its former owner in 2008.

2. *Operational rules*: Three families were created with the Atlas.ti family manager – III. Land registered, IV. Primary registrations as well as Vouchers – and set in a row as a global filter whose respective quotations were edited in a subsequent window (see figure 5-19 below).

(a) In Samegrelo, it was estimated that in one of the villages 95% of the people did not register their land, although some have bought land that had been previously registered (see below). Whereas some believed that 90% of the villagers did not know that they had to register, others claimed that "there are no big lands, everybody knows what belongs to whom". People emphasized that registration – particularly obtaining adequate documentation by getting proof from the archive and then registering free of charge – only becomes a topic of interest when somebody is buying or selling land, as using the fallow land of others is generally possible. Problems especially arose, however, when a buyer claimed "that additional land belonged to the plot that he has bought", since the plots, based on earlier records, have not been geo-referenced. Recognition of adjacent land plots (held in unlawful possession) through attestation of five neighbors and approval of the land commissions was considered very costly at GEL 700/ha.

---

[277] Law (2007, Art. 7[4)]; amended under the Law of Georgia No 3889 of 7 December 2010).

[278] Ordinance No. 231 on the Regulation of Certain Issues Related to the Registration of Titles (2012).

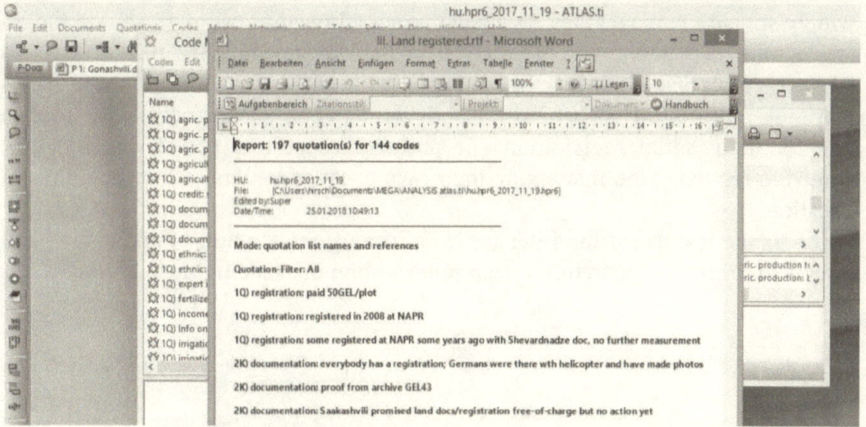

Figure 5-19: Global filter for 'III. Registered land'

Among the participants of a focus group in Kakheti it was believed that most, at least more than half of the villagers did not register their land due to high costs (GEL 700–800 in total), even though it was also mentioned that in 2012 it was announced on TV that it was possible to register during a limited period, i.e. in August and September, free of charge. For this limited-time opportunity, if the name of the actual owner was similar to the one found in the records, no further proof from the archives was necessary but only re-measurement and the attestation of two to three neighbors. "It was easy!", according to one respondent. Another group, however, emphasized the high costs of registering in cases where a family owned many plots or had land in their possession that was recorded under different names. For such cases, documentation from the household book was necessary, which added to the overall costs. Thus, it was said that the majority of the villagers have been waiting for a government program to register free of charge. Some respondents also mentioned that the legal rules were changed in 2012 such that land henceforth has to be surveyed professionally – in contrast to the much cheaper previous practice of using coordinates from Google Maps to register land.[279] Among the respondents of the ethnic Azeri villages, the first group stated that "some registered at NAPR some years ago with the Shevardnadze document, no further measurement" needed. One of the respondents who registered in 2008 mentioned that he had to pay 50 GEL [20.50 USD] to register his plot. The other group of villagers estimated that a maximum 30% of the inhabitants have registered land, as they were also expecting Saakashvili to make land documentation and registration free of charge. Among ethnic Georgians, the first group declared that most people did not register even though information was

[279] According to participants from focus groups conducted in 2013, the practice of surveyors themselves using Google Maps to measure land has been widespread.

widely spread via magazines, by local municipality administrators and on the internet. One respondent who did register in 2013 revealed having paid 50 GEL to register a plot located far from the village. In all of these cases – from interviews that took place in 2013 and 2014 – both farmers and experts recalled a need to pay for their initial registrations. If problems occurred, people unanimously agreed to regulate land matters on their own or let surveyors re-measure plots in question.

(b) As the results for the filter for 'IV. Primary registration', the meaning and effects of primary registrations seem to be ambiguous (see figure 5-20).

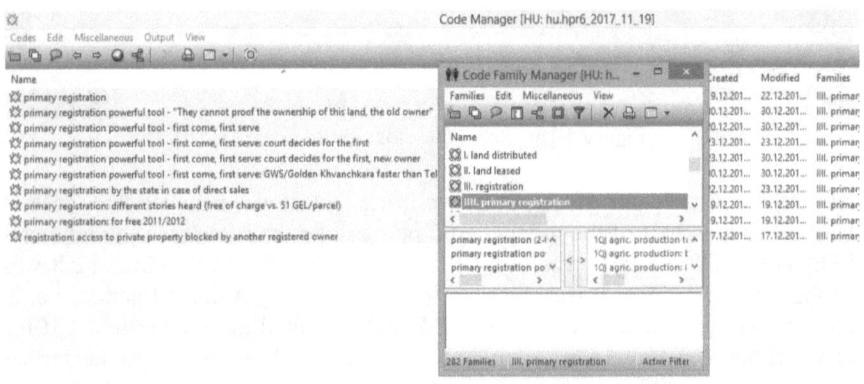

Figure 5-20: Global filter for 'IV. Primary registration'

According to a legal expert, the constitutional court of Georgia generally decides in favor of those who have registered land first. Hence, primary registrations have become a powerful tool leading toward a first-come, first-serve strategy vis-à-vis land for both the state as well as for individuals: As was illustrated above the state registers land as the owner, as occurrences in 2007–2008 have shown, also of land that has been already registered before. (Unfortunately, several attempts by the author to obtain an appointment for an interview with a representative of the legal entity entrusted with the privatization of land in state ownership (MoE) were unsuccessful, and the last cancelation of an appointment by one of the organization's employees on the phone clearly revealed fear about providing information to the author.) Thus, according to a Georgian land tenure expert (and as will be shown below), the government registers land as the owner for that the MoE sells the land to interested (natural or legal) persons by direct sale.[280] On the other hand reports by a legal expert illustrate how disputes of individuals about dubious borderlines of agricultural sites turn out positively for those being legally informed and having

[280] Interview with a land tenure expert, GIZ, Georgia, July 24, 2014, Tbilisi.

the means to register first.[281] Besides, another expert informs about primary registration of land used as alternative source of income by those who speculate on potentially important logistical or agro-ecological sites.[282] Another land tenure specialist report of severals cases where former agricultural land was designated as nature conservation area so that people would loose interest and the state subsequently registered the land.[283]

(c) The voucher program was a very popular means for rendering support to the agricultural sector, commonly perceived as very positive: All farmers reportedly benefited from the vouchers, either by having ploughed and harrowed the land, or by buying fertilizers, machines or any kind of inputs. Whether having land actually registered in the Public Registry did not play any role to benefit from the government's voucher program. But although the program targeted minor landowners holding land of maximum 1.25 ha, people in both ethnic Georgian villages in Kakheti give account of assistance rendered to land owners holding up to 5 ha. Respondents in both ethnic Azeri villages state that only 70–80% of the people's land got tilled and that additionally they were allowed to purchasing inputs: In one village the remaining plots were too small for a tractor to pass through; in the other ethnic Azeri villagers describe that the tractor arrived too late for the soil's cultivation. In contrast, respondents in both ethnic Georgian villages state that all of their land got tilled; moreover, people told to have paid on top to have let more land be cultivated, while others just got the money handed over without having any ground work done. The handling of vouchers in Samegrelo underlines the lack of the villages' arable land resources: The vouchers were only partially used for the land's tillage but people mainly benefited from buying fertilizers and, by adding more money, purchasing machines, e.g. mini-tractors.

Thus, while villagers were happy to get assistance, the process to keeping control of and monitoring the appropriate issuance of vouchers by local administrations (sakrebulo) turned out extremely difficult, as is emphasized by a *gamgebeli,* i.e. head of administration, in one of the ethnic Azeri villages as well as by (some well-informed) respondents in focus groups held in Kakheti and Samegrelo. In line with the aggregated land reserves (as is indicated in people's household books which are recorded in the respective sakrebulo) the government transfers vouchers to the municipalities. Problems appear first of all with view to the fact that the villagers only need to testify the amount of the eligible area covered by a voucher orally, without showing any legal proof of title to the land to be tilled. As a consequence, in a few cases not only the household head, but also his wife and his children applied for vouchers. The administration of vouchers which covered both

[281] Interview with a legal expert, interview on June 21, 2013, Telavi.

[282] Interview with a government representative of the MoE, Department of Urban Planning on August 1, 2014, Tbilisi.

[283] Interview with a wine producer's operation manager, June 25, 2013, Telavi.

tilling as well as purchasing machinery and inputs turned out difficult and very time intensive.

3. *Comparison* of the ways legally and illegally possessed land was registered in Georgia with the results of the process indicates that (a), although the initial law requiring land registration – which mandated loss of right to land titles if not complied with – eventually had a strong impact, for the majority of respondents the rule rather constituted a symbolic act, as the costs of formalizing ownership of plots comprising not more than 1–2 ha were grossly disproportionate to their benefits. (b) The subsequent legal rule on free-of-charge primary registrations was enforced as written, although there were no legal safeguards provided to compile a list of existing rights holders before new registrations were entered into NAPR's database. Consequently, this rule did not end up benefitting small-scale agricultural landholders to finally obtain ownership rights but, rather, served the government's interest to further privatize state-owned land according to a first come, first served strategy. (c) The relationship between legal provisions for benefitting from the governmental voucher program that targeted landholders possessing registered land resources of up to 1.25 ha and its realization is characterized by a high degree of incongruity: first, support was rendered to landholders without requiring any proof of title; second, support was given independent of whether the land had been registered before; and, third, the program also provided assistance to those who possessed up to four times more land than was originally stipulated.

IV. Privatized land

1. *Collective-choice rules*: The next step in the process of privatizing state-owned agricultural land started (a) in 2005 and targeted not only direct sale of leased land (see section II. above) but also public auctioning of non-leased state-owned resources (see chapter 5.3). Privatization through special auction applied to land of former leaseholders who were not interested in acquiring the land and targeted the auctioning of land to local Georgian natural persons and legal entities (Art 3e) held by the Ministry of Economic Development of Georgia (MoE); for land not sold at initial auctions, a second, open auction was held for all Georgian citizens and legal persons of private law registered in Georgia (Art 3f). The law also now covered *type II* land resources that had not been sold before (see above). For the privatization of non-leased agricultural land, a special auction – held only for local villagers (Art. 13) – would be organized. An open auction would "be held only if a special auction arranged in connection with the non-leased land has failed to choose the winner" (Art. 13.1). Art. 2.3 of the law (2005) excludes the privatization of pastures and cattle driveways. Art. 4.1 of the law (2005) declares agricultural land used by the Georgian Apostolic Autocephalous Orthodox Church to be Church property, transferred to the Church's private ownership free

of charge under Art. 4.2.[284] Legal reforms in 2010 introduced alternative privatization procedures via transferring state-owned agricultural land through auctions, direct sale and competitive direct sale (chapter 2, Art. 7.1).[285] An electronic auction to acquire unleased state-owned land may be initiated by a citizen of Georgia or a legal entity (Art. 8.1); information of such sales is available on the website of the MoE.[286] In 2012, the Constitutional Court of Georgia declared legally guaranteed ownership only to Georgian natural persons to be unconstitutional.[287] As the recognition of rights to both lawfully possessed as well as squatted land was legally restricted to natural persons from 2012,[288] the law since then has also included *type III* land reserves which are sold through electronic auction. Though pastures were excluded from privatization by the law in 2005 (Art. 2.3), legal changes in 2010 (Art. 4.1 (bb.a)) allowed for privatization of pasture land leased out before July 30, 2005, thus having a retroactive effect. Cattle driving routes are still kept exempt from privatization.

In 2007, the President of Georgia launched the "100 Agricultural Enterprises" initiative that targeted the privatization of land parcels of more than 50 ha, to be sold to both domestic and foreign investors under specific conditions, such as setting up processing plants on estates acquired under preferential financial terms. These purportedly direct and competitive direct sales of state land were based on a decision of the Government of Georgia, carried out under mandatory investment conditions, and the legal foundations for transferring land through (competitive) direct sale were introduced via legal changes in 2010.[289] According to the literature, the procedures initiated in 2005 targeting direct sale of leased land were often said to be delayed; auctions organized by the local administrations were reportedly biased, seeing that by and large the population did or could not participate due to missing information. Accordingly, legal changes in 2010 charged the MoE with the organization of auctions. Response to the reforms launched for (competitive) direct sales was allegedly low; a prominent example, however, was a direct sale to Ferrero, which at first found cemeteries among its acquired land reserves.

2. *Operational rules:* Now specifically including hazelnut processors and wine producers in my qualitative analysis, here the acquisition of their land resources (if there were any) is analyzed by examining codes summarized by Atlas.ti within a filter called 'study II: land'. In Samegrelo, a Ferrero employee disclosed that the sub-branch was founded in Georgia in 2007 under preferential terms; the land sale to Ferrero at a cheap rate was linked to "the following investment conditions: you

[284] Further free-of-charge transfers through direct sales applied "to citizens of Georgia who lived or still live on occupied territories and remain homeless" (Law 2007, Art. 6.1), based on decisions made by the President of Georgia upon proposal by the MoE.

[285] Law of Georgia on State Property (2010).

[286] See www.eauction.ge; www.nasp.gov.ge.

[287] Law of Georgia on Agricultural Land Ownership (1996).

[288] Law of Georgia on State Property (2010, Chapter 2, Art. 7.1).

[289] Law of Georgia on State Property (2010).

had to invest a certain amount; if you buy this land for that price you have to employ a certain amount of people, and you have to build something (…). We brought hazelnut dryers".[290] In total, the company acquired 4,000 ha of land in Samegrelo, mainly from the state but also from local people who initially were against the launch of Ferrero's plantations in their neighborhood. The purchases were organized step by step in 2007–2008. After the finding of cemeteries and third-party claims to approximately 300 ha of the sites Ferrero just had acquired, the company complained to the government, returned the land and asked the government to replace it. The so-called Shevardnadze documents that Ferrero received as titles to its land were seen very critically by the company as "simply prints, not geo-referenced, fake". Hence, the respondent from Ferrero went himself to the field: "Mainly the land problems solving process is…that you go to the plot to understand the problem, is it occupied or is it just claimed illegally, or you… but however, if there is a significant problem…our main interest, our main target is 'don't go against the local population', refer to the government again and let them solve the problem".[291]

Likewise one of the hazelnut processors interviewed had acquired land from the state and declared similar problems: Of the 160 ha that the firm held in total in 2013, about 20 ha of land bought by auction in 2006 was still pending due to third-party claims, and court cases were still ongoing.[292] In contrast, all three wine producers in Kakheti have land under their ownership that was mainly bought from individuals who had held unregistered land: The first firm, holding 300 ha of land, mainly bought land privately in fragmented, scattered plots (each of 90, 60, 30, 16, 6 and 4.5 ha in size) whose different localities represent specific wine sorts, such as Tsinandali, Alexandrauli, Mujuleturi or Saperavi.[293,294] But since the people selling their land owned several plots of small size, the firm had to negotiate heavily before being able to get the land registered in the Public Registry: In one case, 60 peasants were involved, who altogether held Shevardnadze documents for 170 plots, which totaled only 10 ha in the end. The second comapny held 450 ha of land in 2013; about 230 ha were initially leased and bought in 2005 by direct sale (the main plot is 9 km distant from the firm, the farthest is 32 km away and some is located in-between, about 16 km); more land was subsequently leased and then acquired or bought outright, when available. The third of the three companies bought 120 ha of land in 2008 from individuals as well as from smaller wineries who had to give up their business as a consequence of Russia's ban on Georgian agrarian produce (2006–2013); however, the process took quite long, as

---

[290] Interview with the Head of General Affairs Department at LLC AgriGeorgia/Ferrero Location Georgia Industry on July 8, 2013, Zugdidi.

[291] Ibid, July 8, 2013.

[292] The legal expert explains that "an old owner can go to the court and sue not the new owner but the Public Registry for adopting this decision"; interview on June 21, 2013, Telavi.

[293] Interview with ibid. on June 21, 2013, Telavi.

[294] According to the legal expert, there are 18 appellation wines in Georgia (ibid.).

not much land was available in the region, and finally registering unregistered land plots was time consuming. By 2013, the firm held a 50 ha piece for the appellation Naparauli, plus 20 ha of Kindsmarauli as well as 50 ha of fallow land on which 5 ha of further grapevines are planted annually.

3. *Comparison* shows that the rules intended for the sale of state-owned agricultural land in Georgia appear to be in line with the working rules that have been applied and enforced as written. But considering that the land had initially been registered as state-owned and was subsequently sold without due consideration of any existing rights, the government can be said to have intentionally infringed on some people's rights to own land.

5.2.2.4 Governance of land to be privatized

Following the analytical categories of the IoS framework along the present four *action arenas*, namely (I) *distributed land*, (II) *leased land*, (III) *registered or illegally possessed land,* and (IV) *privatized land*, the following part gives an overview of how organizational arrangements were set up to governing the process of land privatization.

(I) Distributed land

Regulations stipulating the distribution of land were ratified by the Cabinet of Ministers in 1992. For implementing and monitoring the privatization process, a State Committee of Land Resources and Land Reform of Georgia was established (see chapter 4.2), while land commissions, comprised of members of the local administration became in charge of distributing agricultural land, and certificates (ADRs) were issued by sakrebulos beginning in 1992. As findings of the focus groups I conducted have shown (see chapter 5.2.2.1), land was partially distributed under different historical and geographical conditions and, as is known from the literature (see chapter 1.2), under considerable influence from the local elite, meaning members of the local administrations and former kolkhoz and sovkhoz leaders. Moreover, it was reported that ethnic Armenian and Azeri minorities were discriminated against in the allocation process. As also reported, employees of the Ministry of Agriculture and Food Industry (MAFI) tried to block USAID's reform initiatives, leading to new institutional arrangements that resulted in organizational changes which thenceforth, for example, authorized the Ministry of Agriculture and Food (MoA) to collect (statistical) information on land resources.

(II) Leased land

Beginning in 1996, the process of leasing out state-owned agricultural land became a responsibility of the municipalities, which acted through local offices of the State Department for Land Management (SDLM), which was in charge of the management of land resources until 2004 (see chapter 4.3)[295]. At the same time,

[295] Interview with a land tenure expert, GIZ, Georgia, June 15, 2013, Tbilisi.

the Bureau of Technical Inventory was put in charge of surveying and registering apartments and buildings, the Ministry of Environmental Protection for the administration of land resources vis-à-vis the quality of their soils, while the Ministry of Agriculture and Food kept (statistical) information on land resources. Overlapping competencies among the State Department and the other Ministries combined with political influence, control and corrupt practices led to the mismanagement and underfunding of the SDLM. The decision to gain assistance from many donors for the process of formalizing agricultural-land property rights was made within the SDLM, whose leadership endeavoured, according to an expert, to gain as much (financial) support as possible, without being aware of the differences between the technical approaches and procedures of various donors.[296] However, an NAPR government representative emphasized that the NAPR had preferred to *only* apply for an extension the KfW project – in order to minimize dealing with the multiple and even conflicting approaches of different donors – but the GEL 30M [USD 12,149.400] due for the almost ten-year long cooperation with KfW beginning in 2000 was still pending and needed to be repaid by 2039.[297] Due to the many problems plaguing the SDLM, it was eliminated, its cadastral data was transferred to the NAPR between 2004 and 2006, and surveying of land was handed over to the private sector.

(III) Registered or illegally possessed land

Beginning in 1997, USAID and Booz Allen Hamilton were engaged to support privatization of agricultural land, conducted by the subsequently founded Association for the Protection of Landowners' Rights (APLR). The latter was created in 1998 and became the leading figure in formulating and implementing legal rules that targeted land privatization and formalization (see chapter 4.4). Thereby, the organization developed from being an advocate for newly created (small) landowners to a consultancy corporation for large landowners; it also assisted the government's large infrastructural projects, developed on agricultural land previously registered by villagers.[298] Meanwhile, the organization is deeply entangled within political networks:[299] It was established by Vano Merabishvili, who served as its president from 1996–1999; he was then appointed Minister of the Interior from 2004–2012 and named Prime Minister by President Saakashvili for a half-year term until the 2012 October parliamentary election and the ensuing change of government. Moreover, APLR directly participated in the development of land-registration procedures and setting up of the registration system, while its leaders systematically became part of the Georgian government. As mentioned by one

[296] Ibid., June 15, 2013, Tbilisi.

[297] Interview with a government representative of NAPR's International Relations Department on July 12, 2013, Tbilisi.

[298] Interview with a representative of the Ivane Javakhishvili Tbilisi State University, Division of Human Geography, July 24, 2014, Tbilisi.

[299] Interview with a land tenure expert, GIZ, Georgia, June 15, 2013, Tbilisi.

respondent, "for five of them at least, in the Ministry of Economy, Ministry of Justice, even Ministry of Interior, even Prime Minister, he [i.e. Merabishvili] was then... So it was a quite influential think tank".[300] According to some experts, it is a feature of the Georgian political system that a number of NGOs have been closely incorporated into the political processes and become tantamount to being governmental bodies:[301] "So there is an NGO, nothing special, but once their representatives move to the government they just pick up from those NGOs who fulfill the things what they need".[302] Besides, as was emphasized by the same professional, "many former government members have established their own NGOs and now are part of civil society, and [I mean] a lot of them, almost everyone".[303] As he explained, "this is how it works: it is a small society, and people are quite flexible, especially when they had or still are having influence, they know how to transform themselves in order to be a player in the politics and social life of the country".[304] He also emphasized, however, that "an NGO should not be intervening in the things that are purely governmental responsibilities, and the other way around. This is important. But, unfortunately, we don't have such checks and balances quite well established".[305]

Since competencies were moved from SDLM to NAPR in 2004, registration of land has been under the umbrella of the Ministry of Justice (MoJ). Meanwhile, the Ministry of Economy and Sustainable Development (MoE) has been in charge of selling agricultural and non-agricultural state land and urban land planning, whereas the Ministry of Agriculture (MoA) has been responsible for land management, land use planning, monitoring, and alienation of land. Forest management has become the task of the Ministry of Energy and Natural Resources,[306] and the Ministry of Environment Protection has taken charge of national parks and protected areas.[307] Local self-government bodies (sakrebulo) oversee non-agricultural land in coordination with the MoE. However, this governance system – comprised of a number of legal entities, each engaged with duties related to different

---

[300] Interview with a representative of the Ivane Javakhishvili Tbilisi State University, Division of Human Geography, July 24, 2014, Tbilisi.

[301] Stated by several experts in interviews hold in Tbilisi and Telavi during June to July 2013 and July to August 2014.

[302] Interview with a representative of the Ivane Javakhishvili Tbilisi State University, Division of Human Geography, July 24, 2014, Tbilisi.

[303] Ibid., July 24, 2014, Tbilisi.

[304] Ibid.

[305] Ibid.

[306] The Ministry of Energy was merged in December 2017 with the Ministry of Economy and Sustainable Development (MoE), with the latter becoming its legal successor (see http://www.energy.gov.ge/show%20news%20mediacenter.php?id=793&lang=eng).

[307] Likewise, in December 2017, the Ministry of Environment and Natural Resources Protection was combined with the Ministry of Agriculture to form the Ministry of Environmental Protection and Agriculture (see http://moa.gov.ge/En/News/4401). When running an internet search to obtain information from a website for the newly formed governmental body, the search engine directly connects to the MoE (see http://www.moe.gov.ge/en/home).

land reserves but under the control of the MoE at its very top level – lacks synergy and coordination.[308]

The process for compiling the country's land register and maintaining existing records consists of two components: (A) the parcel, represented via a map or plan based on information derived through (i) *demarcation*, (ii) *indication* and (iii) *surveying*, and (B) its owner or registrant, who is seeking (iv) *registration* of state grants (v) *adjudication* of existing rights and/or (vi) *conversion* of deeds (Simpson 1976: 219; see figure 3-4 chapter 3.2.3). In order to trace the genesis of the manner in which the Georgian register and cadaster was compiled, these six elements are used to help structure the following investigation.

(A) The choice for defining an appropriate unit of record depends on the purpose of registration, which could include recording the varying sizes of land plots, their soil quality and hence their (fiscal) value, or grouping severals plots into either a single unit of operation where land planning is concerned (regulatory) or forming a unit of ownership, regardless of whether the land plots in question are contiguously located, to give emphasis to particularities of ownership. By defining land plots according to individual units of use, apparently mainly fiscal motives led to NAPR's guidelines to collect land information plot-wise.[309] (i) Seeing that no national coordinate system had yet been set in place, no *demarcation* of boundaries was realized, but vaguely defined general boundaries were used (see figures 5-21 and 5-22 below); given that both kinds of issued certificates of land ownership, namely ADRs and CLOs, were not (adequately) geo-referenced, (ii) the official positioning of a parcel's boundaries by *indication* was thus only realized after on-site surveying and registration in the Public Registry; as presented above (see chapter 4.4), the (iii) *surveying* techniques used by the two main donors (i.e. USAID and KfW) differed: KfW made use of photogrammetry on household plots and the remaining state-owned land, leading to GPS-based digital maps; thus, only subsequent surveying would provide the final cadastral plot boundaries. Meanwhile, USAID used traditional field-surveying techniques on privatized land plots, resulting in non-georeferenced land parcels and ownership certificates.

[308] Interview with a land tenure expert, GIZ, Georgia, June 15, 2013, Tbilisi.
[309] Regulation No. 509, Art. 1.5 (2011).

Figure 5-21: Example (1) on parcel demarcation

Figure 5-22: Example (2) on demarcation

The primary registration of property, usually conducted by state officials, was initially realized by privately licensed, sub-contracted surveyors hired by the APLR who rendered support to small- and large-scale agricultural land owners via a limited number of surveyance professionals trained by APLR.[310] Later, after having become a self-financed private company, the APLR was involved in the country's large-scale infrastructure projects which resulted in the eviction of people from their titled land, as commission-based surveying activities, carried out "quick and dirty", had resulted in imprecise and vague indications. Moreover, instead of measuring the plots in accordance with the documented property-right indications – primarily the total size of the property, as indicated in the Shevardnadze document – the survey work was done on an empirical basis of what could be seen and, thus, included whatever area was fenced in.[311] As a result of both the vague indications and the conflicting measurement techniques of the two main donors, and with due consideration of the time that had passed until the first registration was even entered into the APLR's database, the NAPR leadership stipulated remeasuring of the ground on-site and keeping the donor organizations' old records to later compare possible changes in ownership over time.

---

[310] It is reported that 700 people benefited from this training, namely "all those surveyors that you today find on the market", according to a land tenure expert, GIZ, Georgia, June 15, 2013, Tbilisi.

[311] Interview with a government representative of NAPR's International Relations Department on July 12, 2013, Tbilisi.

(B) Registration of owners and their respective interest in a particular plot of land – while at the same time requiering to consider any further subordinate or derivative interests affecting land ownership – commenced with (v) free-of-charge systematic *adjudication* led by the APLR (in collaboration with its partners, i.e. Booz Allen Hamilton, USAID and Terra Institute), which ended up only covering a part of the agricultural land reserves in Georgia. After USAID's completion of the LMDP's two project cycles, adjudication was legally changed toward a sporadic adjudication process, while surveying activities were shifted to the private sector. As any existing right of ownership had to be renewed by re-measuring and registration in the newly founded Public Registry by mid-2011, the (vi) *conversion* of deeds, meaning possible proof of title based on ADRs and CLOs, was ignored but subordinated to "deeds" registered thereafter. When primary registration became legally free of charge, the government implemented the (iv) *registration* of state grants, either in potential touristic zones, where ascertaining of existing rights was simply neglected and people lost titles to their formerly registered land or, for non-registered land, where the government shifted ownership rights back to the state.

When discussing the above-described procedures for formalizing property rights in Georgia with experts, a central-government official stated that there were "a lot of reforms in the previous years concerning the registration (…) policies", and a representative of the Ministry of Agriculture confirmed that "laws on that [formalization] changed a lot".[312] However, whereas the previous organizational structure lacked transparency and independence, the more recent organizational set-up is said to provide these features by separating customers' applications from the final registration process,[313] as a means for guaranteeing that, for example, an applicant's ethnicity does not play any role in properly registering her property.[314] Although the management of registrations and the cadaster is centralized, customer service was then decentralized throughout Georgia. In practice, according to one NAPR representative this means that, for sale-purchase agreements for example, "it is not mandatory to go to the notary when you buy your property, you can come directly to our office, or to our authorized users".[315] With this setup, the front offices of NAPR are decentralized, so registering land is possible in many outlets throughout Georgia at NAPR's so-called authorized users, such as banks,

[312] Interview with a government representative of the MoE, Department of Urban Planning on August 1, 2014, Tbilisi; interview with a government representative of the MoA, July 13, 2011, Tbilisi.

[313] Interview with a government representative of NAPR's International Relations Department on July 25, 2014, Tbilisi.

[314] Interview with a representative of the Ivane Javakhishvili Tbilisi State University, Division of Human Geography on July 12, 2013, Tbilisi.

[315] Interview with a government representative of NAPR's International Relations Department on July 25, 2014, Tbilisi. According to the same expert, legal grounds to become one of NAPR's authorized user is a Memorandum of Understanding'.

insurance companies, notaries or surveyors.[316] These organizations work independently of centralized back offices, where applications are examined together with relevant land documentation and, eventually, registered.[317] It is worth noting that, although applications might be dismissed when inaccurate cadastral information is delivered, where for example no surveying has actually taken place – as learned from interviews with farmers, local government representatives and experts – in some cases, "land surveyors just copy/paste some coordinates from the maps, that they had from these old cadastral maps… And it is accurate in the end, but they don't go to the field".[318,319] However, if someone registered in 2004–2005 without having provided any graphic information – as the former registration system was not yet based on a cadaster – the NAPR expert explained that "you can now specify it, apply, and provide a cadastral plan with all these documents, it is free of charge".[320]

The current governance structure for formalizing land ownership is, thus, based on hybrid governance (see chapter 3.1.5), an institutional arrangement that connects many formerly loose decision-making points. This is in line with meeting the criteria for recognizing the boundaries of a local unit mentioned above (see chapter 3.2.2): (i) *control*, (ii) *effciency*, (iii) *political representation*, (iv) *equitable distribution of costs and benefits*, and (v) *self-determination* (Ostrom et al. 1961: 835).

Local units in the present example meet the condition of (i) *control*, in that NAPR clients can obtain the relevant services required for property registration,

---

[316] Interview with a government representative of NAPR's International Relations Department on July 25, 2014, Tbilisi. The NAPR has its main office in Georgia's capital, Tbilisi, in the newly erected Public Service Hall, in addition to ten regional offices; it also "cooperates with more than 350 private entities such as banks, real estate companies, notary offices and others involved in the real property market. These entities have access to the Public Registry's databases and are authorized to receive citizens' applications for registration" (Rolfes, Jr., Leonard & Grout (2013: 4).

[317] Interview with a government representative of NAPR's International Relations Department on July 25, 2014, Tbilisi.

[318] The current practice for measuring land is based on CORS (Continuously Operating Reference Stations), which is said to be "much better than GPS, because the public registry [employees] actually can see whether the land surveyor went on the field or not. They can monitor CORS devices online and can see whether someone actually measured at the location or not", according to a government representative of NAPR's International Relations Department on July 25, 2014, Tbilisi. Moreover, "it is faster, it is more precise, for example whereas GPS devices might have deviations of 10 to 15 meters, CORS has only several centimeters", according to an interview with an expert of TI Georgia on July 25, 2014, Tbilisi.

[319] I also learned that NAPR provided all certified surveyors with the orthophotos produced by KfW which, if no changes occurred, were allowed for the registering of land (interview with a government representative of NAPR's International Relations Department on July 25, 2014, Tbilisi).

[320] Interview with a government representative of NAPR's International Relations Department on July 25, 2014 Tbilisi.

either at one of NAPR's local offices or at authorized users; (ii) the governance system has also created an *efficient* structure in the sense that the system allows for boundary criteria to be changed and adapted, the technology in use is applicable and employees are adequately skilled to facilitate the organization's undertakings. The latter point though was viewed critically by one expert though, seeing that these employed are "new-comers (… who) do not know much about the system, because mostly they are lawyers, and lawyers don't know anything about cadastres".[321] The criterion (iii) *political representation* is met by providing the service on an adequate scale (size and feasible number of administrative units providing the good) to the appropriate target group, here the public, according to decisions made in the common interest and also taking account of informal proceedings. The organization's set up for offering its services, with its administrative split into front and back offices, enables it to operate on an appropriately broad scale by giving authorized users the right to easily and publicly access front offices located throughout the country, as at least banks and notaries are to be found in smaller towns. Political representation founded on unbiased rulings is, in principle, guaranteed by the agency's legal subordination to the Ministry of Justice of Georgia. The short-term authorization of free-of-charge initial registrations for a three-months period in 2012, as announced via television commercials by the former government, was undertaken explains one expert "because (…) it was a pre-election period and the former government decided to register free of charge to get votes, and that is why they cooperated and they got involved".[322] Meanwhile, others paid for initial land registration, illustrating how property registration has been used as a populist measure.

(iv) The *equitable distribution of costs and benefits* has been a sensitive issue, since even though the institutional change toward registering ownership might bring about economic vitality (see chapter 1.2), the restructuring of the governance system involved considerable social and economic costs. (v) As, according to an NAPR representative, the organization is both financially independent and legally competent to take its own decisions, such as splitting up front and back offices or the choice to implement compulsory sporadic registration, the entity's *self-determination* seems to be warranted and financially apt for internalizing the public good. However, considering that the Public Registry only provides private titles to landed property, the organization lacks means for providing institutional arrangements based on home rule for the country's pivotal land resource: pasture land. Empirical results in Georgia "on one of the most celebrated hypothetical cases in the law-and-economics literature" (Ellickson 1986: 624), that is Coase's (1960) famous cattle trespass dispute, show a traditionally rooted "pro-cattleman

---

[321] Interview with a representative of the Ivane Javakhishvili Tbilisi State University, Division of Human Geography, July 12, 2013, Tbilisi.

[322] Interview with ibid. on July 12, 2013, Tbilisi.

'fencing out' rule that many grazing states [in the U.S.] adopted during the nineteenth century" (ibid.: 660).[323] In Georgia, where due to its mountainous nature "the tradition of running cattle [i.e. mainly sheep] at large remained strong" (ibid.: 661), open-range rules apply, with the government being responsible for the maintenance or renewal of adjoining trespassing routes.[324] The assignment of private rights to pasture land though has forced animal breeders to either lease pasture land from individuals who at the time could afford to buy formerly leased land or to graze their animals at home;[325] it is also reported that, in some cases, municipalities have registered pasture land for common use.[326] The focal point closely connected to home rule and self-determination is to successfully internalize a public good, namely both its private and social costs and benefits (Pigou 1920). The individualization of collective grazing land might fuel competition over the country's particularly scarce winter pastures and, hence, generate considerable private and social costs.

As presented above (see chapter 3.2.2), the criteria for comprehensively understanding and assessing the performance of a polycentric governance system consist of examining patterns of (1.) *cooperation*, (2.) *competition* and (3.) *conflict* among the entities working within it – in this case, the Public Registry's governance system (Ostrom et al. 1961).

1. *Cooperation*: The involvement of various actors of the private sector, such as banks, notaries, and surveyors – so-called authorized users – enables many people easily gain access to the organization's services. By decoupling the service for applying for property registration (front office) from its subsequent assessment (back office), this structure impedes the likelihood of opportunistic or discriminatory behavior. However, with regard to general cooperation among all legal entities involved in the present governance structure, the information they generate is hardly shared among each other.[327] Thus, the decoupled organizational structure has its advantages for those seeking to register but, at the same time, evidently allows the government to generate asymmetric information.

2. *Competition:* The government of Georgia set up a fee-based contracting system for covering the costs of property registrations, with a standardized price to

[323] Here, Coase (1960) describes a reversal of liability rules from "property-rule protection of the rancher to liability-rule protection of the trespass victim" (Ellickson 1986: 626). He also provides support for the pro-cattleman rules in the northern states of the U.S.; however, western states in the U.S. also adapted statutes to rule on "specific fence technologies that farmers can employ to revive their rights to recover for trespass damages" (ibid. 1986: 660). Pro-cattleman rules are the legal opposite of "the traditional English rule that the owner of livestock is strictly liable for trespass damage" (ibid. 1986: 660).

[324] Interview with an expert on sheep herding and exporting on July 13, 2011, Tbilisi.

[325] Focus group in Kakheti, July 2, 2013.

[326] Interview with an academic expert on August 10, 2011, Tbilisi.

[327] Land resources are administered under the legal umbrella of the MoJ; further administrative bodies that relate to land resources include the MoE, MoA, Ministry of Energy and Natural Resources, Ministry of Environment Protection as well as local self-government bodies.

be paid by NAPR clients, regardless of whether they apply at one of NAPR's Public Service Halls or at a branch of the Registry's authorized users.[328,329] NAPR clients thus benefit from a quasi-market choice situation: The provisioning of the public good, separated from its production and offered in accordance with specific performance criteria, is therefore not directly subject to rivalry but is, rather, built-in within a self-regulating system.

3. *Conflict resolution* schemes are, according to an NAPR representative, available in cases where an application is rejected: "you can call the hotline (…) and if it is stopped, the registrar also indicates the reason why it was interrupted. And the cititzens have one month for providing all the documents or missing documents or any errors in the documents".[330] If the applicant, having paid the registration fee of GEL 50 in advance, provides the missing information in time – which could include remeasuring their or a neighboring parcel, due to overlapping indications – the application will be successfully entered into the database; if not, the applicant must start the application process again from scratch.[331] As was reported by a legal expert, many court cases followed the introduction of digital registration, mainly due to acquisition of land that was left unregistered.[332] Here, "[a]n old owner" was able to "go to the court and sue not the new owner but the public registry for adopting this decision" regarding their property.[333] Today, however, the chance of former "owners" being able to claim rights of ownership has been reduced to almost zero, according to the expert.[334] According to another expert, many Georgians have not been registering because they are "only slowly understanding the legislation". As the "legislation becomes more regular, people will understand and approach the law [to register]. … With time, that will be solved, step by step".[335] But, as was said during one of the focus groups in Zugdidi, "[p]eople's main motivation to register is if somebody else wants to buy their land. Otherwise, usually they can't see the need".[336] This situation arises because,

[328] Regulation No. 509, Art. 2a. According to the representative of NAPR's International Relations Department, "[f]or notaries and state authorities this service is free of charge. We give this authorization access free of charge. But others…the private sector are paying for this authorization, annually. For example, surveying or real estate companies pay an annual fee of 500GEL; banks 1,000GEL. But for others it is free", interview on July 12, 2013, Tbilisi.

[329] It is interesting to note that "Public Service Hall" is translated into Georgian as "House of Justice" (*iustitsiis sachli*).

[330] Interview with a representative of NAPR on July 25, 2014, Tbilisi.

[331] Interview ibid. on July 25, 2014, Tbilisi.

[332] Interview with a legal expert on June 21, 2013, Telavi.

[333] Interview ibid. on June 21, 2013, Telavi.

[334] Ibid.

[335] Interview with a representative of a former Deputy Agriculture Component Leader, EPI (USAID), then FAO, Georgia, on August 4, 2014, Tbilisi.

[336] Focus group in Zugdidi, July 8, 2013. According to findings from the focus groups, further factors that led people to register land include campaigns of NGOs, observing neighboring villages with land-related problems, local gamgebeli who motivated local citizens (in the

according to a land tenure expert, "neither the surveyor, NAPR, nor any other governmental agency is responsible for the measurement's accuracy" but, instead, the sanguine potential land owner.[337] As expressed unanimously by the experts, the reason for this confusion stems from the lack of legal requirements for surveying techniques and cadstral standards. On the other hand, as was described by an NAPR representative, the Public Registry consists of "two databases within the registration database: a textual database, with textual information on the property, and a graphic cadastral database. They are linked to each other with a cadastral code".[338] However, as was stated by a land tenure expert, those being responsible at the NAPR, "[t]hey think that technology is the solution to everything. Technology is nothing unless the individual, or a group of individuals, do not [...] use the technology properly. So, they introduced this technology [...] so fast that they probably thought much less about the purpose of the technology and how to use it".[339]

The governance structure of the organization for formalizing property rights, comprised of public and private entities, is thus built on a hybrid structure but is not especially marked by polycentricity. The provisioning of the public good, separated from its production and offered in accordance with specific performance criteria, is not directly subject to rivalry but is, instead, ensconced within a self-regulating system. The system of conflict resolution, though, is questionable, seeing that in cases where an application is rejected, the registration fee paid in advance becomes a sunk cost that has to be invested again for a new application. As applications can be rejected due to overlapping measurements, the initial registration information entered into the NAPR database is given extraordinary power to block ensuing entries, although no further litigation mechanism is set in place to mediate the problem.

(VI) Privatized land

Legal changes in 2005 paved the way for the privatization of state-owned agricultural land, first, to primary local leaseholders; second, to local legal entities; and, finally, to the public (see chapter 4.3). Essentially, lessees enjoyed a right of preemption and preferential payment schemes that enabled acquisition up to 2011. Acquisition of the remaining unused (vacant) state-owned agricultural land was first opened to physical persons and legal entities registered in Georgia through special auctions. At this point, such land could only be sold on the initative of a Georgian natural or legal person through open auction by the local sakrebulos,

---

case of the two Azeri villages). Meanwhile, reasons not to register were mainly associated with financial obstacles.

337 Interview with a land tenure expert, GIZ, Georgia, July 24, 2014, Tbilisi.

338 Interview with a representative of NAPR's International Relations Department on July 12, 2013, Tbilisi.

339 Interview with a representative of a former Deputy Agriculture Component Leader, EPI (USAID), then FAO, Georgia, on August 4, 2014, Tbilisi.

while the MoE prepared a purchasing deed which served as proof of title to be registered at the NAPR. In 2007, the government initiated direct sale of large-scale agricultural land parcels through the MoE in cooperation with the APLR and the MoA. The privatization of adjacent land plots based on local land-commission testimony about land being held in unlawful possession, likewise beginning in 2007, was characterized by an expert "as 'a kind of amnesty' for people occupying land [...] although not many took advantage during these times".[340] From 2011, the leasing of agricultural land became prohibited, whereas land could be acquired through on-line auction or by direct sale from the MoE.

Pasture land, in leasehold prior to July 30, 2005, became subject to privatization in 2010 if the lessee had purchased the land before May 2011. Since then, the remaining pastures have been leased out directly or through auction by the MoE. As explained by a legal expert, the MoE announces public auctions on its website but not locally:[341]

> If the MoE wants to sell some land, it consults with the MoA on individual parcels, although the MoA is not obligated officially to participate in this process. But the document comes and the attached suggestion is as such: 'Would you mind if we try to sell this parcel, if you think it is agriculturally important, or selling this parcel in this way wouldn't harm our' ... and things like that. And the MoA provides some consultancy in this regard. But again, it is entirely the prerogative of the MoE to sell this land plot. That's how it operates now.[342]

Consequently, a hazelnut producer from Zugdidi said that he needed to constantly monitor the site for about a year until he found a suitable piece of land.[343]

The agricultural land governance structure in general is marked by a division of competencies among many legal entities, with the Ministry of Economy and Sustainable Development of Georgia, which is in charge of privatizing land under state ownership, playing a predominant role. The governance structure for the organization for formalizing property rights, comprised of public and private entities, is built on a hybrid structure but less marked by polycentricity than is typical for such structures.

5.2.3 Evaluating the outcomes

The evaluation of outcomes – the assessment of costs under alternative institutional arrangements – is the final step taken in the present analysis. The distribution of costs within the context of Georgian agricultural land privatization is estimated here according to the degree to which people's entitlement to land has been protected over time (see chapter 3.1.4). The initial assignment of rights has been

[340] Interview ibid. on August 4, 2014.
[341] Interview with a legal expert on June 21, 2013, Telavi.
[342] Interview with a representative of a former Deputy Agriculture Component Leader, EPI (USAID), then FAO, Georgia, on August 4, 2014, Tbilisi.
[343] Interview with a hazelnut producer on July 8, 2013, Zugdidi.

decisive with respect to gaining legal status, because entitlement based on either property or liability rules disparately impacts the ex ante or ex post timing of interference, who has to commence bargaining, who has to bear the costs of the transaction and litigation, and, thus, sets incentives for the tendency of interference. A property rule is an entitlement that provides right-holders with the highest degree of security (i.e. from interference from others and, if so, not to bear any related costs), whereas entitlement based on a liability rule make intrusion more likely and protecting people's rights costly for them.[344] The present section thus traces the relational character of people's agricultural land ownership rights over time in Georgia.

Regulations enacted by the Cabinet of Ministers granted eligible households a (claim) right to possess a specific amount of *type I* land, distributed by a *property rule*,[345] which was then given legal capacity by law based on a decision of Parliament in 1996.[346] However, the majority of new land owners were reportedly left without any proof of title, and former records in the archives on people's land use during the Soviet era were ignored. At the same time, the Parliament enacted a law requiring the registration of ownership via a legal entity which had not been formally set up yet.[347] As a result, with the introduction of these two inconsistent principles, those who did have proof of title lost their relevant qualification again, while those without a title could hence only claim recipient rights, that is 'imperfect' claim rights, as no specific duty bearer could be specified. People's former fully protected entitlement was in this manner changed into a *liability rule*, which left both those with and without titles, even though to different degrees, vulnerable to the actions of others, to wit giving others the privilege to intervene, while the responsibility for bargaining ex post and bearing the costs of stopping intervention rested with the actual right-holders. By the end of the 1990s, new ownership certificates were fabricated and issued systematically by USAID together with surveying done by newly trained surveyors – a process that was marked again by both inaccuracy and insufficient distribution to the new land owners.[348] In 2007, President Saakashvili authorized a sporadic registration process;[349] the associated surveying work though, after having merged the dissimilar data produced by the various donors, did not actually result in people gaining proof of title and was then only used by NAPR registrars to track the historical record of land plot boundaries for later registrations. A new law in 2007 required the actual right-holder with

---

[344] Entitlement via an inalienability rule prevents any transactions from taking place (see chapter 3.1.4).

[345] Resolution No. 48 (1992); Resolution No. 128 (1992); Decree No. 503 (1993).

[346] Law of Georgia On Agricultural Land Ownership (1996).

[347] Law on Land (Immovable Property) Registration (1996).

[348] Presidential Decree No. 327 (1999).

[349] Law of Georgia on Recognising Property Rights Under the Possession (Ownership) of Physical and Private Legal Entities (2007).

proof of title ex ante to bargaining among neighbors to compromise on land demarcations; for the ensuing creation of a new measurement plan and registration at the Public Registry, costs were outsourced and shifted solely onto right-holders. In contrast, ownership rights to land "unlawfully" used that adjoined titled land were legitimized by local land commissions, namely the very same people who were in charge of distributing land in the beginning of the 1990s. By the end of 2010, President Saakashvili had authorized that those with the appropriate legal standing to prove land ownership to be given a last chance during a one-year time window to avert losing their claim right by finally registering their land in the Public Registry.[350] After that, the former right-holders lost their (claim) right but remained with no-right and, hence, were in equal standing with other Georgian citizens who had the privilege to acquire land through electronic auction.[351] Further legal changes in 2012 then set primary registration and registration for specifying the area of plots free of charge:[352] Since the government did not ascertain existing rights when compiling the Public Registry, on the one hand, these legal changes provided the government with the power to alter probable existing rights-relationships, including former rights to ownership; on the other hand, however, the situation favors those who registered land in the early days when the NAPR had been established but had not compiled cadastral data yet (2004–2008).

Type II land resources were leased beginning in 1996 via a *property rule*.[353] The allocation process, though, once more favored the well-informed and influential (local) government officials. Conveying use into ownership rights to agricultural land (except pasture land) became possible in 2005 and was highly supported by price reductions and payment schemes.[354] For this, the government granted entitlement based on a *property rule*. As the NAPR had not begun compiling cadastral records yet, these new land owners were moreover exempted from paying any registration fee for specifying their plots.[355] In 2010, the ex post facto law that allowed acquisition of pasture land (leased out before July 30, 2005) again favored well-informed and influential circles.[356] As sheep grazing routes became exempted from privatization, the costs for their maintenance were shifted to the public.

The implications of these developments include the following: Whereas the average citizen who benefited from land distribution was forced to watch enforce-

---

[350] Law of Georgia on Public Registry (2008); Law of Georgia on State Property (2010).

[351] Law of Georgia No 3889 (2010).

[352] Regulation No. 509 (2011); Ordinance No. 231 on the Regulation of Certain Issues Related to the Registration of Titles (2012).

[353] Law on Agricultural Land Leasing (1996).

[354] Law on State-Owned Agricultural Land Privatization (2005).

[355] Ordinance No. 231 on the Regulation of Certain Issues Related to the Registration of Titles (2012).

[356] Law of Georgia on State Property (2010).

ment of their newly gained rights in continual decline, at the same time, they experienced steadily rising costs for protecting their land. In contrast, well-informed and influential circles enjoyed the highest degree of protection and, thus, paid the least for benefits gained, namely at first vis-à-vis flows from land as a resource and, eventually, from future benefit streams stemming from holding land as an asset.

5.2.4 Summary

Based on the Institutions of Sustainability (IoS) framework, this chapter focuses on the transaction as its unit of analysis, transaction properties and the (mis)alignment of governance structures. In particular, it examines, first, the way land was distributed throughout Georgia; second, the rights of minorities during the process of receiving land; and, third, on the effects of the "wholesale" approach to land distribution in contrast to the "incremental" market approach to property rights formalization.

Land, which comprises the unique physical features of being immovable and "everlasting", unites social functions related to both land as an asset, based on its renewable resource stock, and as a flow, which may either generate yields or rent. The properties of land-related transactions are, consequently, greatly determined by asset-specificity, particularly by site-related factors (e.g. quality of soils, roads, or vicinity of major markets) nature-related attributes (e.g. climate, floods) as well as by social artifacts, i.e. time and scale which affect the degree of benefiting from future (financial) streams. In the present case of the Republic of Georgia, technical aspects have also played a significant role, for compiling a public registry and cadaster consists of a set of complex interdependent transactions that are linked to considerable set-up costs; especially social costs have been incurred, as the institution of private property rights is novel in Georgia, where contractual relationships have traditionally been based on oral agreements and reputation effects. Due to the dual nature of right relationships, the formalization of property rights is highly marked by jointness, coherence and complexity. Frequency has played a decisive role in Georgia, as related to the historically new and one-time initial assignment of property rights to land, which has impacted distribution of future costs and benefits among the members of Georgian society.

The interviews conducted for this study reveal that, first, land has been allocated according to different criteria throughout the country. In addition to the general procedure of allocating land to primary rural households, the Republic's first President, Gamsakhurdia, allowed restitution and granted households in parts of Western Georgia a six-month time window to reclaim land which had belonged to their ancestors before the Soviet era. Furthermore, land was allocated in some cases in both Samegrelo and Kakheti according to village size and available land resources, whereas some individuals were reportedly allowed to pay money to receive more land. Though it has been said that only a small part of the population obtained proof of title during these first years, the present study shows that the vast majority of respondents hold proof of title, even though some had to pay for it whereas others obtained it free of charge.

Second, regarding the local conditions of the ethnic Azeri population in Kakheti, people generally work one or two plots, with a total of up to 0.5–1 ha on average; only a few lease additional land (those holding large-sized plots of land were, with the exception of one individual, absent among the respondents interviewed). Land possessed in ethnic Azeri-populated villages in Kakheti seemed to be less than in ethnically Georgian-dominated sites, but more data would be necessary to prove this assumption. In Kakheti, few of the respondents lease state-owned land, and using a part of other villagers' land against payment in kind was the usual and most-stated answer given by about a third of focus-group participants. The majority, however, expressed that they were suffering from the limited amount of available land resources in both ethnic Azeri- as well as Georgian-populated sites. Samegrelo seems to be rich in individually possessed areas that lie fallow but lacks "reserve land" to be leased from the state. In Western Georgia, hilly terrain has led to the formation of typically dispersed settlements where orchards, hazelnuts and corn grow on typically yellow soils in a subtropical climate on or near people's household plots, and cropland is generally located further away. Meanwhile, cows, goats or chickens are kept at home, for pasture land is rather scarce in this region. The majority of the land is occupied by perennials, as arable land is hardly to be found. Agrarian production in Samegrelo benefits from a favorable climate, where particularly hazelnuts, citrus, bay leave, kiwis and fennel belong to the main cash crops. In contrast, the relatively flat terrain in Kakheti is more densely populated, with plots arranged in rather linear settlement types. The household plot is usually smaller in the East, but additionally possessed parcels of crop- and perennial land tend to be larger. The area – marked by high-quality soils that range due to intra-regional climate differences from black to brown and middle-brown – provide the ground for extensive farming of grapes and grains; also fruits, in particular peaches, grow in large parts of the area. Given that pasture land is abundant, the region is home to traditional sheep herding. Today, agrarian production is concentrated on one or maximal two hectares on average. Seeing that off-farm employment is limited, subsistence farming is still widespread; thus, agriculture is family-related and mainly based on mutual help among neighbors.

Regarding, third, the effects of the "wholesale" approach to land distribution in contrast to the "incremental" market approach to property rights formalization, my findings show that both procedures – to wit, the initial assignment of rights to agricultural land – generated increased uncertainty rather than contributing towards spurring agrarian production. In the first case of wholesale distribution, statutory provisions for distributing land were enforced differently, not only from region to region but also from site to site. The results show that land, both to be distributed and later to be leased, was assigned by influential political figures and allocated according to social norms dominating in Georgia, that is the preferential treatment of people of the same origin, without providing any proof of title within the first years of the reforms. But, the provision of titles was also characterized by irregularities under President Shevardnadze, seeing that differences occurred concerning whether people had to pay for a title or got it free of charge. Moreover, whereas the vast majority of respondents in the field confirmed having received a title during this period, the contrary had been indicated by public officials and those donor agencies in charge of implementing reforms. It thus seems that these reports were used by public officials to justify further legal changes that ended up perpetuating people's insecure legal status. The wholesale approach to land distribution benefited local officials and those close to the government, while the process for most of the rural population fostered perpetual legal uncertainty and ran counter to positively affecting agrarian production.

Meanwhile, the incremental market approach to formalizing property rights with the help of USAID illustrates how the Georgian government under President Saakashvili preserved the status quo of land allocated and succeeded in coopting the newly found Association for the Protection of Landowners' Rights (APLR). Political leaders instrumentalized the APLR to eventually gain control over potentially significant land reserves by not only shifting costs disproportionally to claimants for securing ownership rights to land, while using free-of-charge registrations as a popular means – and cynical ploy – for gaining support just before the 2013 presidential elections. Constantly modifying the legal rules and leeway, culminating in the adoption of laws with retroactive effect which eventually targeted the privatization of pasture land, gave insiders the chance to snatch a valuable chunk of a resource that once served the entire community.

6 CONCLUSIONS AND RECOMMENDATIONS

Strictly speaking registration is a device to give certainty to the ownership of land and to enable dealings to be conducted quickly and cheaply; it is not intended to change tenure but only to ascertain and give form to existing tenure so that current practices can more readily be regulated and organized.
(Simpson 1976: 227)

This chapter presents the conclusions derived from my empirical and theoretical analysis of the process of agricultural land privatization in the Republic of Georgia as well as a discussion of its limits and implications. The first part focuses on the main empirical findings and conclusions that I have derived from quantitative analysis as well qualitative fieldwork. After that, theoretical conclusions are drawn that seek to contribute to the law and economics literature in general and regarding privatization, with the formalization of property rights as its last step, in particular. Third, political implications are depicted that form the basis for further recommendations, and the last part outlines topics for further research.

6.1 Empirical conclusions

Institutions give structure to everyday life, reduce uncertainty and set incentives. As such, institutions are created or evolve over time and continuously change, as do the economy and choice sets of social actors. Institutional arrangements evolve to facilitate exchange in society, give structure and help its members to form expectations. Thus, property rights in particular are understood as policy instruments which delineate actor choice sets, meaning for example who is protected by a right from interference or who has to initiate bargaining to change a situation in light of costly and asymmetric information. Formalization of property rights assumes a change in relative factor prices, signals scarcity, and thus a need to assign exclusive, unambiguous property rights which allow using property as collateral and, hence, gaining access to credit. In contrast, the quantitative study of Georgian privatization of agricultural property rights presented here reveals that financial means were the decisive factor for rural dwellers when deciding whether to register their property. Those who chose not to register rely on agricultural production internalized within their family and generally do not appear interested in achieving potentially higher income in agriculture by cooperating with others. Instead, their being involved in the village community seems to substitute for seeking sanctuary via "secured" land titles. Thus, my qualitative study suggests that the outcomes from the Georgian land reform process can be primarily understood in terms of theoretical propositions and empirical findings regarding a transition-specific feature of institutional change, namely *uncertainty*. An average farmer is not only faced with uncertain institutional preconditions in Georgia's rural hinterland (urban bias), but frequent legal changes have made estimates about possible outcomes difficult have ended up providing people with choice sets marked by a high discount rate that tends to give rise to myopic behavior. In particular, my analysis illustrates a strong interplay between formal rules and informal constraints (and opportunities) that in Georgia led – with the help of the international donor community – to an amalgam which constantly produce uncertainty for the average constituent and has overwhelmingly benefitted those belonging or close to (former) government circles. This was achieved by continuously changing the legal rules of the game and, with view to non-informed sheep breeders, eventually enacting a law with retroactive effects, thus preventing any reasonable way to plan for the future; modifying and thus undermining people's proclaimed ownership rights, protected initally by a property rule but then reduced to a liability rule; setting up governance structures and services to be so costly that, instead of being able to gain access to credit, most people were not able to afford to re-measure and register each plot of their land and, thus, lost their legitimate claim to its title; and privatizing pasture land and in so doing causing an artificial shortage of the limited, formerly collectively used land resource, thus resulting in further tightening the choice set available to the average rural citizen.

The *pace of institutional change* in Georgia – where market transactions primarily based on custom and informal ties were replaced by processes of liberalization, with the aim of penetrating the market – has been marked by both abrupt and gradual elements. Distribution of land allocated for both private ownership as well as leasehold via a "wholesale" approach initially pushed by the government came as an unsurprising and abrupt course of action, leading to insider trading and benefiting those who were part of or close to influential local government circles. In contrast, the initial assignment of the corresponding rights to land that had been distributed proceeded gradually, eventually sporadically, and in the end remained incomplete, as many of the newly assigned land owners were left without proof of title and legislation was set in place which did not match existing organizational requirements. The implementation of legal rules stipulating prior surveying and registration in the Public Registry for gaining ownership inaugurated a change from people having their right to title protected by a property rule to a less-secure liability rule. Hence, the situation worsened for those unable to afford the appropriate means to re-measure and register any of the land plots they possessed; meanwhile, influential government officials reportedly grabbed land located in areas envisioned for large infrastructural projects, even at the expense of eligible land owners who had registered the land. However, the assignment of ownership rights to land in leasehold – the majority of which is reported to have been kept for economic or speculative purposes by those belonging or close to local government officials – proceeded steadily, and since then the associated titles granted have been protected by a property rule. The retroactive alteration of constitutional provisions which previously exempted pasture land from being privatized represents a fundamental disruption of the status quo at the expense of the average farmer, who is dependent on common pasture land. Institutional changes induced in collaboration with the international donor community to promote land-market transactions were implemented gradually. New institutional arrangements, such as introduction of the Public Registry or auctioning of the remaining state-owned land, were adopted by means of regularization which – viewed from the outside – seemed to provide a stable institutional environment for the marketization of land. Nevertheless, via processes of situational adjustment, influential government members (and/or those closely related) took advantage of any reform carried out by manipulating rules (e.g. as reflected in regional differences in land distributed); by redefining relationships in their favor, (e.g. occupying organizational entities entrusted with land formalization, including advocacy groups and surveyors); by maneuvering between donors, resulting interim in the nesting of multiple and parallel institutional (foremost technical) systems; as well as by constantly changing the rules, which culminated in the adoption of laws with retroactive effect. Thus, whereas abrupt reform steps were used ad hoc to secure future benefit streams from the *flow* of Georiga's land resources, the interest of these elite actors in land as an *asset* was transferred and secured in a series of incremental steps.

Seen from the perspective of land administration, four aspects of how the Georgian reforms were carried out seem highly critical and are discussed below, to wit, on the one hand, (i) *sporadic adjudication* and (ii) the *individualization* of customary (common) land tenure to achieve land registration as well as, on the other hand, vis-à-vis the cadastre, (iii) the *land surveying sector* in general and (iv) the unit of use as the *unit of record* in particular.

(i) The *adjudication* of existing rights proceeded sporadically – a policy decision which, from a theoretical point of view, seems hard to justify. As a cardinal principle, adjudication is not an instrument for changing or creating new rights. Yet, this is exactly what members of the Georgian government have used it for. Rights to previously transferred state grants have never been conveyed with absolute assurance and, as a result, existing rights were not ascertained but simply ignored and now depend on people's (ability and) willingness to pay. Further, systematic adjudication raises public awareness and, consequently, tends to serve as an effective safeguard against rent-seeking, whereas a sporadic approach favors insider circles. The same applies from a more technical point of view, as the systematic approach facilitates surveying work via economies of scale and scope and, hence, saves costs in the long run. But, above all, a systematic approach enables updating of outdated survey data, such as that created by the international donor community, while simultaneously simplifying re-planning, if measurement, adjudication and conflict resolution are conducted in parallel. In this manner, registration can be accomplished from the bottom up, area-wise. By contrast, the sporadic approach adopted in Georgia has shifted costs from the public purse to the individual and prevented the realization of a participatory process for compiling the registry and completing registration in the near future. The situation has been aggravated by the fact that, as will be emphasized below, the NAPR has refused applications that indicate overlapping plot boundaries. Hence, the first person who registers their land defines (with the respective proof of evidence) the property boundaries and might, thereby, be in a position to block adjoining neighbors' applications from registering their land.

(ii) As the quantitative analysis has shown, the vast majority of the villagers is engaged in small-scale animal husbandry. Privatization of formerly common pasture land thus led to an artificial shortage of village grazing land, which has exacerbated the situation for rural dwellers while supporting a concentration of wealth among those who were more informed or belonged to the upper social layers (so-called home rule neglected). Likewise, from a technical point of view, the use of a modern land information system would not necessarily require *individualization* of all types of agricultural land but could, rather, facilitate registering village pastures community- or village-wise. Considering that animal husbandry represents an integral part of the country's cultural heritage, in a region where more than half of the land is comprised of pasture, a combination of diverse governance forms that support a variety of classifications of pasture land within the Puplic Registry could be an alternative to privatization, one which would save social costs, on the one hand, and serve economic purposes on the other. In practical terms, this would mean preserving communal pasture land in areas where grazing land is scarce. The argument is supported by the fact that pasture routes, linking winter and summer pastures, were not subject to privatization but have been continuously maintained by the state. This incongruency suggest that certain private interests have enjoyed preferential treatment at the expense of the public good.

(iii) The *land surveying sector*, comprised of a number of surveyors who were trained in the course of the USAID project, is not subject to any legal regulations. Specification of surveying and measurement techniques, in the form of norms and standards with the aim of creating an impeccable system of land records, has been missing since its inception. In fact, by and large the surveyors who benefited from USAID training were not only engaged in the regular process of surveying but were also involved in illegal appropriation of land located in economically promising zones and are, according to interviews, today the main operators in their field. At the same time, no other educational or regulatory means – as well as standards thereof – have been set in place by the state to promote and train new surveying staff. The condition of needing to re-measure plots via use of global positioning system (GPS) techniques is a standard was only imposed by the Public Registry – not by law –after e.g. communities in ethnic Azeri-inhabited villages applied for land registration based on dimensions they had taken from the internet (Google Maps). As no ground methods had been used by surveyors to produce the plot specifications, the applications were refused. This process displays two particular features of the governance system employed for formalizing property rights in Georgia. As mentioned above, it was not the government, through laws and regulations, that has defined the organizational rules and procedures for recognizing and ascertaining initial land rights according to clear norms and standards but, rather, the NAPR, under the umbrella of the Ministry of Justice set the rules. This has given the NAPR considerable *power*, specifically the ability to establish new legal relations (and liabilities) among legal subjects (and objects) who have to obey these rules. (At the same time, though rules regarding property registration already existed, those who grabbed previously registered land enjoyed *immunity*, i.e. not being affected by whatever subsequent legal relation was set up and were not held liable to pay compensation.). Thus, althoug the initial ownership structure should not be changed by the formalization of property rights, the first-come, first-serve approach benefits individual interests and neglects the development of a comprehensive, community-based process for ascertaining land titles.

(iv) Finally, the *unit of record*, which defines the link between an owner and a given plot of land, depends on the purposes for which a cadaster is being used. In Georgia, the unit of record has been defined by the 'unit of use', which lays emphasis on land values assigned to individual parcels of different sizes for the purpose of taxation. The decision to adopt this unit – especially as it affected the initial ascertaining of land rights – is questionable, seeing that an average land owner in Georgia is in possession of two to three parcels and has, hence, been obliged to re-measure and register several plots individually. Alternative forms for ascertaining and registering several land plots under one owner include using a 'unit of operation' that makes up a farm, which underlines land use, or to form a 'unit of ownership' to highlight the particulars of land ownership. Given that plots are scattered, as land has been distributed according to its local availability (Lerman 2004b) – a particularity of the Georgian land distribution process – both of the just-mentioned options present ways that could have more easily and equitably dealt with the problem of scattered plots. The unit of operation comprises two or more units of use that belong to the same owner and make up a farm; the focus here is on land use and development, which is subject to specific regulations or enjoys particular support. Meanwhile, a unit of ownership comprises two or more units of operation, consisting of different farms that are possibly under distinctly different governance structures (e.g. leasehold). Registering multiple units not only minimizes the sunk costs for initial registration in the short term but also facilitates taking account of Georgia's different land use types – animal husbandry (pasture land), orchards (perennials) and agriculture (arable land) – which are often combined within an individual household. Against the background that more than half of Georgia's population relies on subsistence farming, the procedure established during the reform period has run counter to the principles of establishing a system of titles, especially simplicity and cheapness, which are key for earning the population's acceptance of such a new form of governance. In contrast, the costs for ascertaining initial land rights have been constantly shifted to the formerly proclaimed land owners who, if they cannot pay the piper, are at risk of losing their claim rights. (Moreover, in some cases the formalization of property rights has been used as a political instrument that promoted the re-nationalization of previously distributed land for re-sale). The concept of land ownership created for the average farmer who has no influential ties has thus been greatly marked by indeterminacy, which sets incentives for myopic economic behavior and detrimentally affects agricultural production.

6.2 Methodological conclusion

Based on the results described in this thesis, application of grounded theory in the course of an abductive inquiry, combined with exploration of quantitative and qualitative data appears to be a fruitful approach. First, grounded theory offered a flexible research process: though built on the researcher's prior insights, it still leaves scope for elaborating new ideas. Second, initial quantitative examination of the research field – integrated into a study on a closely related research subject, namely the study on the role of social capital in rural areas in Georgia (CSS 2012) – provided important reference points and exhibited trends that served as background information that helped to contextualize the topic. Third, the results of the quantitative study were able to be tested and explored in-depth qualitatively and, if necessary, falsified. For example, initial results indicating extraordinarily high registration rates in primarily ethnic Azeri-inhabited areas were, as was revealed subsequently due to the qualitative study, the result of a common endeavor by villagers to register their land, using Google Maps as the basis for measurement, instead of GPS-based ground methods, at a time where standards for surveying and measurements had not been legally defined yet. These applications were, according to an expert interview, most probably refused by the NAPR, which then required re-measurement, the feasibility of which depended on farmers' willingness to pay in light of their respective budget constraints. Combining quantitative and qualitative data, the mixed-method approach proved fruitful for understanding how the institutional framework for the Georgian land-privatization process was changed and enabled evaluation of its outcomes.

6.3 Theoretical conclusions

By re-examining the process of land privatization in Georgia, in particular the outcomes of property rights formalization via the theories of institutional economics, the present study makes a number of theoretical contributions to the field. By focusing on institutional frameworks and contractual relationships, institutional economics provide an analytical lens that facilitates highlighting the costs of exchange and moving the nature of rights relationships to the fore, which for the present study has clearly shown who the winners and losers of the Georgian reforms have been, and why. In line with the literature on public choice, my analysis supports the view that politicians and administrators have been exposed to a structure of perverse incentives, which have allowed them to favor particular constituents (Ostrom & Ostrom 1971), or been affected by interest groups in the political process (Hagedorn 1991, 1994). Employing the Bloomington school's conceptualization of public goods, with its emphasis on decentralization and polycentric systems, for my evaluation of the organizational structure established for the individualization of property rights according to normative criteria (control, efficiency, political representation, equitable distribution of costs and benefits, and self-determination) as well as performance criteria (cooperation, competition and conflict) proved to be valuable for illustrating the effective functioning of the NAPR's hybrid organization (Ostrom et al. 1961; Ménard 2004a). For example, my analysis shows that the distribution of costs and benefits for the Georgian reform has been a sensitive issue, as the restructuring of the governance system involved considerable social and economic costs, primarily at the expense of the most vulnerable. I have also demonstrated that the organization's self-determination is apparently warranted but lacks institutional arrangements based on home rule, seeing that the country's key land resource (i.e. pasture land) became (partly) subject to individualization and, hence, led to an artificial shortage of grazing land used by local communities. Thus, I hold that both the hypothesis as well as the underlying assumptions of the Evolutionary Theory of Land Rights (Platteau 1996) should be rejected, as the formalization of property rights over agricultural land in Georgia has rather artificially throttled the supply of land and given influential individuals the chance to snatch a significant share of this valuable resource. Moreover, as land possessed in rural Georgia is typically highly fragmented, the vast majority inhabitants there can neither benefit from using their land as collateral nor are they able to pay for its re-measurement and registration. Additionally, empirical results from Georgia relevant to "one of the most celebrated hypothetical cases in the law-and-economics literature" (Ellickson 1986: 624), that is Coase's (1960) famous cattle trespass dispute, reveal a traditionally rooted "pro-cattleman 'fencing out' rule" (Ellickson 1986: 660). Due to its mountainous nature, "the tradition of running cattle [sheep] at large remain[s] strong" (ibid. 1986: 661) and open-range rules apply; at the same time, however, the government is liable for the maintenance or renewal of adjoining trespassing routes. Thus, while the newly assigned owners of pasture land may now exclusively enjoy the land's future benefit streams, they are not held responsible for the costs stemming from this institutional change. The Institutions of Sustainability (IoS) framework (Hagedorn 2005, 2008), with the transaction as the unit of analysis, has provided a useful lens for revealing the costs of spill-over effects of the segregative institutions established during the Georgian reforms. My examination of the properties of the transactions established via the reforms has shown that the resulting decomposed structure of the land governance system has evidently allowed the government to generate asymmetric information, which might once again provide the conditions for rent-seeking and opportunistic behavior. For examining the course of and effects stemming from the abrupt and incrementally induced institutional changes in the Georgian case (Streeck & Thelen 2005), particularly Lund (2006) theoretical differentiation between processes of regularization and processes of situational adjustment has proven fruitful for illustrating how (local) members of the government have used sudden legal changes to physically appropriate land, whereas incremental steps have helped them to steadily transfer the corresponding rights and, by the reassignment of liability, shift the costs from the public purse to average citizens at unproportionally high social costs.

6.4 Political conclusions

The political conclusions to be drawn from this study are ambiguous, as they need to be split into two distinct questions: 'what was to be done?' and 'what was politically feasible?' As to the first, the primary objective should have been to compile a register by granting free-of-charge primary registration, village programs to adjudicate and consolidate land, provision of formerly common grazing ground, compensation payments to those who unlawfully lost their land, as well as reforms to develop the surveying sector. In contrast, the results of my study rather suggest that compilation of the register has not been in line with the former government's policy objectives in that, first, systematic registration was explicitly opposed by the former government from the start; second, no information has been given regarding the actual amount of land privatized or leased out; third, the introduction of free-of-charge primary registration was used by the government to re-nationalize (and re-privatize) land reserves; fourth, several attempts by the author to obtain an appointment for an interview with a representative of the legal entity entrusted with the privatization of land in state ownership (MoE) were unsuccessful, and the last cancelation of an appointment by one of the organization's employees on the phone clearly revealed fear about providing information to the author. Hence, the 'land question' still seems to be a quite sensitive issue.

As individualization of ownership in Georgia was put on the agenda by the international donor community, the lessons learned for provisioning of foreign aid for formalizing property rights are in general related to the need to unite the country's individualistic undertakings and harmonize them from the beginning. In the present case, all donor projects had the same goal – surveying to create a registry linked to cadastral maps – but used diverse techniques that produced different formats which had to be mutually adjusted before they could be used and, thus, became a procedure tantamount to a collosal waste of time and financial resources at immense social cost. Viewed within a broader context, my analysis shows that the assignment of exclusive rights after the demise of the Soviet Union – a time during which Cold War narratives declared non-private property to be a Communist bad – was to be realized at all costs in a self-proclaimed 'quick and dirty' process that ended up being at the expense of the average Georgian farmer. Land reforms in the Republic of Georgia – a country of high geo-strategic importance, situated between the Russian Federation in the north and the countries of the Middle East further south – thus served as a political instrument of the Western donor countries to induce a capitalist mode of production based on exclusively assigned land ownership rights. As a consequence, no institutional arrangements were put in place to register commonly used grazing ground under a common property regime and, hence, take into account the country's cultural heritage of nomadic husbandry. Individualization proceeded according to a blueprint designed via a top-down approach, which resulted in a muddle of institutional multiplicity that allowed some influential individuals to exploit the situation in their favor. As many of the formerly proclaimed land owners had been working on land with no adequate proof of title and no financial means to get it registered, a holistic, integrative course of action was required for adjudicating (and possibly consolidating) and assigning private rights as well as providing communal use rights for common grazing ground in a community-based, participatory bottom-up process.

With regard to consolidation, land fragmentation in Georgia has played an important role. In some ways, fragmentation is generally considered a detriment to agricultural production, as plots are on average small and dispersed throughout a region. On the other hand, the results from Kakheti show that different soil types and topography among fragmented plots can enable cultivation of distinct wine varieties and, thus, allow some land owners to take advantage of the country's prominent feature (i.e. site specificity) in the form of an immense variety of scarce but high-quality types of land. However, considering the outcomes of the land reforms, it seems that distribution of non-contiguous dispersed plots served in the past as a safeguard against large-scale appropriation of the land. As is illustrated by Dahlman (1980), a change in ownership or production patterns under such conditions is typically marked by high decision-making costs (Buchanan & Tullock 1962), which inhibits more toward consolidation and, thus, potential concentration of land in the (invisible) hands of a few. Encouraging change of village ownership structures is, thus, questionable and should only be promoted with the utmost care.

6.5 Implications on future research

The Georgian government ran a free-of-charge initial registration program from August 2016 through January 1, 2019. But, as figures from 2018 show, "70% of [the] population, both rural and urban, has not [yet] registered property rights on their land plots" (Business & Finance Consulting 2018: 2). As the quantitative study presented here has shown, the 30% who did register their land are less connected to and involved in everyday village life compared to the majority of the respondents who did not register their land, who share a traditional way of village life, with information spread among neighbors and voluntary provision of help in their neighborhoods. For the latter, communal infrastructure might substitute for the expected gains stemming from property registration. However, it is more likely that such farmers cannot afford to pay for the required re-measurement of their plots. Moreover, as the Public Registry refuses applications indicating overlap with prior registrations, farmers might end up stuck by neighbors who registered too much adjoining land, a situation which requires neighbors to bargain about the exact land distribution before a final re-measurement of the plots. Without a systematic surveying and adjudication process, the objective of eventually compiling the register has been questionable. Future empirical research might be able to provide a more-detailed overview of and ways to deal with these failed efforts to formalize property: Will other donors join the process again to complete the task, or will it become possible to publicly finance an integrative procedure? The recent introduction of a Blockchain Land Registry, meaning an enhanced e-governance system for property registration that can help in registering property and maintaining the register, might improve transparency and generate financial means, but presents a drop in the ocean for supporting the conveyancing of land rights in Georgia. Further research on the individualization of property rights in other former Soviet countries might reveal more satisfactory outcomes, to wit showing ways government has supported a compulsory, systematic registration process or presenting alternative ways of realizing compulsory elements in land registration via other sources of financial support (e.g. microcredit) for bearing the costs necessary to compile a register.

From a theoretical point of view, my research has been based on the Institutions of Sustainability (IoS) framework, with an explicit focus on transactions and their properties, which enabled systematic examination of ex-ante institutional arrangements and their subsequent changes and effects on agricultural production in the course of the land reforms. Regarding the need to support other methods of registration and possibly consolidation of people's land, use of the Institutional Analysis and Development (IAD) framework might yield helpful results on community attributes and rules-in-use which may help to promote and encourage these endeavors. As the quantitative findings presented here have disclosed, the registration numbers in ethnic Azeri-populated villages were significantly higher than those in ethnic Georgian-inhabited areas, while the qualitative results have revealed that Azeri communities agreed to carry out their villages' land distribution by jointly measuring (though use of the internet and Google Maps) and applying for their land's registration (which were most probably refused by the Public Registry for not being appropriate proof of title). Game theoretic techniques can be used to complement my findings on how to affect the bargaining power of players in cooperative as well as non-cooperative settings. Further studies are needed to find an equitable solution for completing the registration process, especially with regard to its costs, how to organize and finance the bargaining and measurement process for registering remaining land with lower social costs, such as through microcredit, public finance (taxes) or foreign aid. As women without any other household members were observed (three individuals from the overall study population) to be living in devastating conditions, assistance should focus especially on the rural poor and women.

From a Political Economy perspective, it should be emphasized that Georgia's integration within the global economy through free-trade agreements and a rising number of super- and hypermarkets that are competing against Georgia's domestic agricultural production has put considerable pressure on the country's rural economy. The rising success of local wine and hazelnut production has, as my study illustrates, been based in both sectors on public–private partnerships that jointly set up educational institutions: Schuchmann Wines and the Georgian government in the form of vocational training at the local technical college and higher education programs in Kakheti's Agrarian University; meanwhile, USAID and Ferrero offer training for farmers to increase the quality of hazelnuts, while offering credit to expand their production. Evaluation of the education and training sector in Georgia might reveal valuable findings for applying such efforts to other sectors of the (rural) economy as well as for comparing the results and their implications with those of similar public–private partnerships in other former Soviet states. From a domestic point of view, the role of intermediaries, both in the provision of agricultural inputs as well as in the acquisition and sale of local agrarian produce, needs to be emphasized. These contractors not only provide local infrastructure in the marketization of local agrarian production but provide credit and trade on the basis of barter transactions, rather than money. An analysis of these contractual relationships might bring to light important results for understanding their implications for local economies, especially with regard to prices or variety of products. According to an agrarian expert, a lack of professional skills and knowledge of agricultural operations has led to a steady overuse of conventional fertilizers and pesticides. At present, the organic sector in Georgia only has a modest but rising share in the agrarian economy. The promotion of public–private partnerships in this very sector, primarily with regard to dairy, cereals or legumes, might lend support to both rural populations and Georgia's ecological systems. Moreover, with increasing temperatures and occurrence of extreme-weather phenomena resulting from climate change, land will become more scarce in Georgia. Thus, site-specific studies on the effects of climate change need to be undertaken and compared with worldwide efforts seeking to help support the resilience and adaptation of local socio-ecological systems.

6.6 Summary

The present study's objective has been to understand the effects of agricultural land privatization in Georgia, guided by its main research questions regarding how political reforms targeting land privatization affected land tenure and land governance and what their effects on agricultural production have been since the break-up of the Soviet Union. In particular, have sought here to grasp (i) whether and how the reforms created winners and losers (ii); whether and how people opposed change in the past or tried to coalesce against it; and, finally, (iii) how the (non-)allocation of property rights has affected agricultural production and, hence, the allocation of wealth among the rural populace. This thesis is a contribution towards understanding the outcomes of an induced institutional change from paper-based land governance arrangements to tenure formalization and its impacts on agricultural production in Georgia. The focus here has been both practical and theoretical, shifting attention from the views of policy makers, donor agencies, legal advisors, and scientists to the concrete outcomes of land allocation and tenure for the rural populace, notably scrutinizing the final process of property registration in Georgia.

Chapter One recounts how my own interest in the topic was aroused by findings from an explorative research visit in July 2011, which was initially directed toward analyzing the problem of Georgia's overgrazed pasture land and rapid increase of exported sheep. Further evidence shifted my focus to property rights to land, the influence of donor agencies and diverse patterns of land governance. While multiple cases of property eviction became publicly known, Georgia ranks first in property registration worldwide (World Bank 2015a). The task of my further research hence became to analyze the different reform steps that have led to this outcome. To explore this problem, relevant New Institutional Economics (NIE) and Public Choice theories are presented that take account of contractual relationships and institutions, the "rules of the game" which govern transactions, and the ensuing distribution of costs and benefits. A review of previous studies on the topic shows that academic literature dealing with land privatization in Georgia is scarce, and institutional analysis in this field has not been carried out yet. Finally, I lay out the research questions, objectives and working hypothesis used to critically scrutinize both the means and outcomes of the reforms.

Chapter Two provides an overview of the tools of the game, to wit the research design and methods used, which are rooted in an abductive research approach and applied together with Grounded Theory and data- and mixed-methods triangulation. With its emphasis on processing data by assembling and discovering features and relationships, the abductive approach is, in line with Grounded Theory, considered apt for analyzing data across different disciplines. By applying data- and across-method triangulation, my study builds on multiple sources of evidence to increase the validity of my approach to the empirical data examined. In addition to regular updating of sources and including secondary literature, the study is primarily based on the findings of a quantitative survey by the Center for Social Studies (CSS) in 2012, which preceded my own qualitative fieldwork in 2013 and 2014. Quantitative analysis of the survey data, based on stratitfied (quota) sampling in villages throughout Georgia and analyzed by the use of SPSS, identifies the general attributes of the villagers and the socio-economic features that might have had an influence on their formalizing of property rights to their land. A subsequent qualitative analysis is carried out to comprehend the reasons why specific features have emerged and, in particular to (1) study the evolution of the Georgian legal framework for land acquisition, (2) analyze the relationships between formal rules and (local) working rules, and (3) evaluate the outcomes of the reforms via their impacts on the ground. Using a case study design, respondents engaged in two agrarian sectors, namely wine and hazelnut, were queried via expert interviews and focus groups. Breakdown and analysis of the qualitative results was realized by use of Atlas.ti (5.2.2), which enables data processing in accordance with the concepts of Grounded Theory.

Chapter Three is centered on presenting the theoretical foundations required for changing the rules of the game, specifically for land privatization, which is understood here as induced institutional change of property regimes from state to private ownership. Differences between defining institutions such as norms and conventions, working rules and property relations are worked out, giving emphasis to the latter as entitlement regimes and examining their implications for the distribution of costs and benefits in contractual relationships. According to the legal correlates defined by Hohfeld (1913, 1917), one person's government protection might be turned into government interference and another's government protection by altering liability in legal relations from a property rule to a liability rule, which essentially affects who has to initiate bargaining and has to bear the costs for re-negotiation and litigation, while also determining the general probability of interference.

The next part deals with governance of nature-related transactions and is centered on property rights regimes and the concept of transaction costs (Hagedorn 2008). Against the background of the recently introduced Public Registry in Georgia, organizational forms are discussed with special emphasis on polycentric systems for the provision of public goods (Ostrom et al. 1961) as well as land administration and registration of ownership (Simpson 1976). At its best, polycentric governance divides the production and provision of a public good so as to secure representation and availability to the outside world, while also offering incentive mechanisms for generating a degree of competition internally to hold costs low. The process of property rights formalization is comprised of two parts, related 1) to determining land plot boundaries (*demarcation, indication, surveying*) and recording them in a map and 2) to legally identifying owners (*registration* of state grants, *adjudication* of existing rights, *conversion* of deeds) and recording them in the land register (Simpson 1976). Property rights formalization is viewed as a form of administrative reorganization and, thus, represents a policy tool – constituting a means, not an end in itself – for achieving carefully defined objectives within an overall public policy framework (Bromley 2008).

For the analysis of institutional change, it is therefore important to differentiate between the pace of the process (abrupt/incremental) and its outcomes (continuous/discontinuous). Seeing that land privatization in Georgia marks a substantial transformation in property regimes from state to private ownership within a rather traditional social context, I have examined it with reference to path dependency. Here, instead of viewing social phenomena through the lens of historical determinism, two complementary processes may generate different outcomes (Lund 2006). Through the process of regularization, institutions and their organizational forms can be created that are both durable and make future events foreseeable. On the other hand, through processes of situational adjustment, people may manipulate newly implemented rules vis-à-vis local working rules and manoeuvre between them in ways that might result in institutional incongruence and ambiguous outcomes. Since Platteau (1996) argues in line with the property rights school of institutional change in his Evolutionary Theory of Land Rights that assignment of exclusive rights to land – when faced with rising scarcity – is warranted, the main ideas of the three most prominent theories of institutional change are reviewed. In contrast, I hold that institutional change should be understood as an endogenous response – in accordance with individuals' and organizations' maximization efforts, bargaining power and ability to extract resources – to exogenous changes which originate from interactions between institutions and organizations competing for economic advantages within contexts of scarcity. The analytical foundations for assess institutional change in the case-study context is provided by the Institutions of Sustainability (IoS) framework (Hagedorn 2008), with its focus on transactions embedded within socio-ecological systems, for understanding the underlying institutional and organizational arrangements involved.

Chapter Four provides an overview of Georgia's land reforms (1992–2015), depicting how the transition from a centrally planned to a market economy was a drastic and far-reaching event from which its populace has yet to recover. Although Georgian agriculture played an important role within the Soviet production and distribution system by supplying wine, tea and citrus, the share of agriculture within the overall economy today has dramatically shrunk. Nevertheless, as the average land possessed by rural dwellers is about 1–2 ha in size, agriculture still plays a central role for meeting the subsistence needs of about half of the population and safeguarding their livelihoods. The first years of independence were marked by inter-ethnic conflicts, civil war and deterioration of the economy due to the break up of the interdependent Soviet and production system. In need of financial means, the Georgian government followed a pro-Western course that entailed a drastic rise in foreign debt, foreign assistance and humanitarian aid. Restitution was avoided because Georgian society had still been organized along feudal lines before becoming the Georgian Soviet Socialist Republic. Land redistribution was chosen a means for avoiding famine by enabling subsistence agriculture. By 1998, about 60% of the agrarian land had been privatized, which was approximately 15% more land than had been foreseen. Moreover, the land allocated lacked any legal guarantees. In 1996 the Government initiated a market-oriented reform by leasing out agricultural land for 49 years. However, as evidence presented here reveals, this process though has been characterized as being biased and corrupted by influential government officials, who obtained land for economic (sub-renting) or speculative purposes. Newly enacted legal regulations by the new government in 2005 aimed to privatize leased as well as unused, vacant land by providing leaseholders with preemptive rights to acquire ownership rights from the state within a five-year period, together with preferential payment schemes over a ten-year term. This process was also marked by fraud, created by financial and organizational hurdles that limited and complicated the process for lessees to acquire ownership rights. In 2010, the government retransferred communal ownership rights for pasture land back to the state and, thus, opened up a window for the acquisition of pastures by setting regulations with retroactive effects. Pastures leased for more than five years became subject to privatization and had to be acquired within about a year. The remaining pasture land was leased out or auctioned by the Ministry of Economy and Sustainable Development. The final outcome of these efforts, namely how much arable land is now under state ownership (again), is not yet known. Popular initiatives for the acquisition of land by a small number of Boer and Punjabi farmers was met by resistance from the local population, who eventually filed a petition to stop land sales to foreigners. As such sales were subsequently ruled to not be in line with the Georgian Constitution, they were suspended in 2014.

Reforms to formalize ownership commenced in the mid-1990s, initially under the influence of German and then, later, American and other international donors. As revealed in this chapter, instead of cooperating, the work of the various donors was marked by competition and ended up producing incompatible data that needed to be harmonized before it could be used for the newly established digital land management system for property registration and compiling a cadaster under the authority of the National Agency for Property Registrations (NAPR), which began processing registrations in 2004 and launched the cadastre in 2006. Prior to that, several state bodies had been in charge of land management and land administration, which led to overlapping, and even conflicting, competencies. The initial decision for systematic registration was altered in 2007 toward sporadic registration. Legal changes in 2012 led to free-of-charge primary registration at the NAPR, which was used by the government to register unregistered land and privatize it. A key development documented here is how inaccuracies in the "quick and dirty" process of land measurement during the early reform period have slowed down the process of compiling the national register. Applications indicating overlapping measurements with other plots are refused by the NAPR, which then demands re-measurement. Consequently, the costs for registering land, and potentially re-measuring it beforehand, have been relatively high and step-by-step shifted to the formerly proclaimed new land owners of the initial redistribution.

Chapter Five presents my analytical results, to wit the findings of the quantitative survey on the characteristics of Georgian land tenure and agricultural production, institutional analysis of (the properties of) land-related transactions as well as my qualitative field work. The findings of the survey reveal distinctive results regarding those who have registered their land and those who have not. Those who have registered land have in general higher incomes and tend not to be native Georgians but rather Muslims. Seeing that they are less limited financially, they believe in self-governance, promote unpaid work, acknowledge rules for collecting money for the poor. Although they are generally focused on internal family matters and less involved in village life – hence, perceiving little change in their villages – they are prepared to work jointly with others to potentially increase their incomes from agricultural production. Meanwhile, those who have not registered tend to be native Georgians, typically Christian Orthodox, with lower income levels. Although they are convinced that the government should assume more responsibility for solving village problems, they are also actively involved in village life, perceive collective action to be (rather) vital in their village, are more involved in political discussions, feel social pressure from commonly held views within their village and are also more sensitive vis-à-vis change there. Additionally, such villagers prefer agricultural production internalized within the family than potentially achieving higher incomes through agriculture by cooperating with others; it seems that being 2involved in the village community is seen as a substitute for seeking sanctuary via "secured" (though in practice precarious) land titles.

Based on the Institutions of Sustainability (IoS) framework, the chapter further focuses on examining the land-related transactions, their properties and the ensuing alignment of governance structures involved in the Georgian reform case. The properties of land-related transactions are strongly determined by *asset-specificity*, particularly by site-related aspects (e.g. quality of soils, roads, vicinity of major markets) nature-related attributes (e.g. climate, extreme weather events) as well as by social artifacts, including time and scale, which affect the degree of benefits that can be derived from future (financial) streams. In the Georgian case, technical aspects have also played a significant role, for compiling a public registry (and cadaster) comprises a set of complex interdependent transactions that are linked to considerable set-up costs. Social costs have also been incurred, as the institution of private property rights is a novel development in the history of Georgia, where contractual relationships had typically been based on oral agreements and reputation effects. Due to the dual nature of right relationships, the formalization of property rights is highly marked by *jointness*, *coherence* and *complexity*. For the Georgian case, in the initial assignment of property rights to land – which was in effect the distribution of future costs and benefit streams – *frequency* played a dominant role, as it only happened once. The top-down policy process for privatizing land pursued in Georgia represents a bundle of collective choice rules that have affected rules-in-use (organizational rules), conceptualized here as a complex chain of actions divided into four action situations according to the primary transaction type: (I) *distributed land*, (II) *leased land*, (III) *registered or illegally possessed land.* My analysis reveals that the rights to distributed land were first protected by a property rule that was subsequently, with the introduction of property registration, converted to a liability rule. In contrast, leased land was, from the first granting of use rights to the assignment of ownership rights, constantly protected by a property rule. Thus, while the vast majority of the formerly proclaimed land owners experienced a steady decline of their legal certainty, former lessees enjoyed constant provision of legal security.

My comparison of the effects of the "wholesale" approach to land distribution in contrast with the "incremental" market approach to property rights formalization illustrates that both procedures generated increased uncertainty rather than contributing towards spurring agrarian production. With the wholesale method, statutory provisions for distributing land were enforced differently, not only from region to region but also from site to site. The results presented here show that land, both to be distributed and later to be leased, was assigned to influential political figures and allocated according to social norms dominating in Georgia, meaning here preferential treatment for people of the same origin. Also, under President Shevardnadze, the provision of titles was characterized by irregularities, especially in terms of the financial costs incurred when some people had to pay for their titles while others got them free of charge. Furthermore, whereas the vast majority of respondents on the ground claimed to have received a title during this period, the contrary had been reported by public officials who, it seems, sought to justify legal changes that would maintain people's insecure legal status rather than improve it. The wholesale approach to land distribution has thus benefited local officials and those close to the government, whereas for most of the rural population it has fostered perpetual legal uncertainty and run counter to positively affecting agrarian production. Meanwhile, the incremental market approach to formalizing property rights, supported by USAID, illustrates how the Georgian government under President Saakashvili preserved the status quo of the initially allocated land and succeeded in coopting the newly found Association for the Protection of Landowners' Rights (APLR). Political leaders instrumentalized the organization to eventually gain control over potentially significant land reserves and shift costs disproportionally to claimants for securing ownership rights to land. My analysis also reveals that, by constantly modifying the legal leeway of new rules, culminating in the adoption of laws with retroactive effect which eventually targeted the privatization of pasture land, lawmakers gave insiders the chance to snatch a significant share of a resource that once served a community.

According to my analysis, the Georgian governance structure in general is marked by a division of competencies among many legal entities, with the Ministry of Economy and Sustainable Development playing a dominant role by registering unregistered land and selling it. At the same time, comprised of public and private entities, the governance structure of the NAPR, the organization set up for formalizing property rights, is built on a hybrid structure that is relatively low in polycentricity. A positive development has been achieved here by decomposing the service for applying to register property (front office) from its subsequent assessment (back office), which is intended to impede the likelihood of opportunistic or discriminatory behavior. In addition, NAPR clients benefit from quasi-market choice: Provisioning of the public good (land), separated from its production and offered in accordance with specific performance criteria, is not subject to rivalry but, rather, operates as a self-regulating system. The scheme of conflict resolution in place, however, is questionable, seeing that rejected applicants must pay the registration fee again in order to reapply. As applications are often rejected due to overlapping plot-boundary measurements, the first plot registered into NAPR's database is given extraordinary power to block ensuing entries and, as of yet, no further litigation mechanism has been set in place.

7 References

Agrawal A (2001): Common property institutions and sustainable governance of resources. World Development, Vol. 29, No. 10: 1649-1672; **29**: 1649–1672.

Akerlof GA (1970): The Market for "Lemons": Quality Uncertainty and the Market Mechanism. The Quarterly Journal of Economics **84**.

Alchian AA (1950): Uncertainty, Evolution, and Economic Theory. Journal of Political Economy **58**: 211–221.

Alchian AA (1965): Some Economics of Property. Il Politico **30**: 816–829.

Alekseeva L, Fitzpatrick CA & Helsinki Watch (1990): Nyeformaly: Civil Society in the USSR. United States: Helsinki Watch Committee.

Alesina A & Dollar D (2000): Who Gives Foreign Aid to Whom and Why? Journal of Economic Growth **5**: 33–63.

Ali DA, Deininger K & Goldstein MP (2011): Environmental and Gender Impacts of Land Tenure Regularization in Africa: Pilot Evidence from Rwanda. Available from: http://www-wds.worldbank.org/external/default/WDSContentServer/IW3P/IB/2011/08/18/000158349_20110818104704/Rendered/PDF/WPS5765.pdf.

Anderson TL & Hill PJ (1975): The Evolution of Property Rights: A Study of the American West. The Journal of Law and Economics **18**: 163–179.

Anderson TL & Hill PJ (1983): Privatizing the Commons: An Improvement? Southern Economic Journal **50**: 438–450.

Apallas AK (2016): The Life Cycle of a Wine Grape: From Planting to Harvest to Bottle. WineCoolerDirect.com, July 6, 2016. Available from: https://learn.winecoolerdirect.com/life-cycle-of-a-wine-grape/.

APLR (2011): Privatization Prices/Land Taxes: New Privatization Prices for Agricultural Land and New Annual Rates of Property Taxes on Agricultural Lands Enforced in Georgia as from January 2011. Available from: http://www.aplr.org/?lang=eng&id=377.

Arrow K (1951): Social Choice and Individual Values. New Haven: Yale University Press.

Arrow K (2000): Economic Transition: Speed and Scope. Journal of Institutional and Theoretical Economics **156**: 9–18.

Arthur WB (1988): Self-reinforcing mechanisms in economics. In: Anderson PW, Arrow K & Pines D (eds.) The Economy as an Evolving Complex System. New York: The Perseus Books Group.

Arthur WB (1989): Competing Technologies, Increasing Returns, and Lock-In by Historical Events. The Economic Journal **99**: 116–131.

Arthur WB (1994): Inductive Reasoning and Bounded Rationality. American Economic Review **84**: 406–411.

Atwood DA (1990): Land registration in Africa: The impact on agricultural production. World Development **18**: 659–671.

Banerjee AV, Gertler PJ & Ghatak M (2002): Empowerment and Efficiency: Tenancy Reform in West Bengal. Journal of Political Economy **110**: 239–280.

Barrell R, Holland D & Pain N (2000): Openness, Integration and Transition: Prospects and Policies for Economics in Transition. Paper prepared for the International Economics Study Group 25th Annual Conference, Isle of Thorns, Sussex, September 2000. Available from: https://www.researchgate.net/profile/Nigel_Pain/publication/5200514_Openness_Integration_and_Transition_Prospects_and_Policies_for_Economies_in_Transition/links/09e4150a3603602f27000000.pdf.

Barrows R & Roth M (1989): Land Tenure and Investment in African Agriculture: Theory and Evidence. LTC Paper 139, Land Tenure Center, University Madison-Wisconsin. Available from: https://minds.wisconsin.edu/bitstream/handle/1793/34120/139.pdf?.1.

Bator FM (1958): The Anatomy of Market Failure. The Quarterly Journal of Economics **72**: 351–379.

Bechtolsheim M von, Graf Hoyos C & Schindler G (2012): Cadastre and Registration Systems in Georgia and Azerbaijan - Experiences from two successful cases: Paper presented at the "Annual Conference on Land and Poverty", the World Bank, Washington DC, April 23-26, 2012.

Becker LC (1980): Property Rights: Philosophical Foundations. Boston, London and Henley: Routledge & Kegan Paul.

Beckmann V & Hagedorn K (eds.) (2007): Understanding Agricultural Transition: Institutional Change and Economic Performance in a Comparative Perspective. Institutional Change in Agriculture and Natural Resources, 26. Aachen: Shaker Verlag.

Bedoshvili D (2008): Elements of a National Strategy for Management and Use of Plant Genetic Resources in Georgia. Sustainable Agriculture in Central Asia and the Caucasus, Vol. 4. Available from: http://www.fao.org/docrep/013/i1500e/Georgia.pdf.

Ben-Porath Y (1980): The F-Connection: Families, Friends, and Firms and the Organization of Exchange. Population and Development Review **6**: 1–30.

Bezemer, DJ/Davis, JR (2003): The Rural Non-Farm Economy in Georgia: Overview of Findings. NRI Report N° 2729. Available from: http://papers.ssrn.com/sol3/Delivery.cfm/SSRN_ID695268_code459916.pdf?abstractid=695268&mirid=1.

Bischof H (1995): Georgien: Gefahren für die Staatlichkeit. Studie zur Aussenpolitik, Nr. 68. Bonn: Friedrich-Ebert-Stiftung.

Black D (1948): On the Rationale of Group Decision-making. Journal of Political Economy **56**: 23–34.

Black D (1958): The Theory of Committees and Elections: Cambridge University Press.

Blanchard O (1997): The economics of post-communist transition. Clarendon lectures in economics. Oxford: Clarendon Press.

Blaug M (1980): The Methodology of Economics. Cambridge: Cambridge University Press.

Bledsoe D (2006): Can Land Titling and Registration Reduce Poverty? In: Bruce JW, Giovarelli R, Rolfes, Jr., Leonard, Bledsoe D & Mitchell RM (eds.) Land law reform: Achieving development policy objectives. Law, justice, and development. Washington, D.C.: World Bank 143–174.

Bluashvili A & Sukhanskaya N (2015): Country Report: Georgia. The Fund Georgian Center for Agribusiness Development. Available from: www.agricistrade.eu/wp-content/.../Agricistrade_Georgia.pdf.

Boettke PJ & Marciano A (2015): The past, present and future of Virginia Political Economy. Public Choice **163**: 53–65.

Bogaerts T & Zevenbergen J (2001): Cadastral systems — alternatives. Computers, Environment and Urban Systems **25**: 325–337.

Böhm A (2004): Theoretical Coding: Text analysis in Grounded Theory. In: Flick U, Kardorff Ev & Steinke I (eds.) A companion to qualitative research. London: Sage 270–275.

Boone C (2007): Property and constitutional order: Land tenure reform and the future of the African state. African Affairs **106**: 557–586.

Boone P (1996): Politics and the effectiveness of foreign aid. European Economic Review **40**: 289–329.

Boucher SR, Barham BL & Carter MR (2005): The Impact of "Market-Friendly" Reforms on Credit and Land Markets in Honduras and Nicaragua. World Development **33**.

Braund D (1994): Georgia in antiquity: A history of Colchis and Transcaucasian Iberia, 550 BC-AD 562. Oxford, Oxford, New York: Clarendon Press; Oxford University Press.

Brazenor C, Ogleby C & Williamson I (1999): The spatial dimension of aboriginal land tenure: Paper presented at the 6th South East Asian Surveyors Congress, Fremantle, 1 – 6 November 1999. Available from: http://citeseerx.ist.psu.edu/viewdoc/download?doi=10.1.1.40.2720&rep=rep1&type=pdf.

Bromley DW (1989b): Economic Interests and Institutions: The conceptual foundations of public policy. Oxford: Blackwell.

Bromley DW (1989a): Property relations and economic development: The other land reform. World Development **17**: 867–877.

Bromley DW (1978): Property Rules, Liability Rules, and Environmental Economics. Journal of Economic Issues **XII**.

Bromley DW (ed.) (1992): Making the commons work: Theory, practice, and policy. San Francisco, Calif.: ICS Press.

Bromley DW (1997): Rethinking Markets. American Journal of Agricultural Economics **79**: 1383–1393.

Bromley DW (2005): Sufficient reason: Volitional pragmatism and the meaning of economic institutions. Lecture presented at the China Center for Economic Research, Beijing University, December 6, 2005. Available from: http://old.ccer.edu.cn/download/5899-1.pdf.

Bromley DW (2006): Sufficient Reason: Volitional Pragmatism and the Meaning of Economic Institutions. Princeton: Princeton University Press.

Bromley DW (2008a): Chapter 10: Land and Economic Development: New Institutional Arrangements. In: Cornia GC & Riddell J (eds.) Toward a Vision of Land in 2015: International Perspectives. Cambridge, Massachusetts: Lincoln Institute of Land Policy 217–236.

Bromley DW (2008b): Formalising property relations in the developing world: The wrong prescription for the wrong malady. Land Use Policy **26**: 20–27.

Bromley DW & Yao Y (2007): Understanding institutional change: Modeling China's economic transformation since 1978 (draft). Presented at the meetings of the Allied Social Sciences Associations, Chicago, January 6, 2007.

Bruce JW (2006): Reform of Land Law in the Context of World Bank Lending. In: Bruce JW, Giovarelli R, Rolfes, Jr., Leonard, Bledsoe D & Mitchell RM (eds.) Land law reform: Achieving development policy objectives. Law, justice, and development. Washington, D.C.: World Bank 11–64.

Bruce JW & Fortmann L (1989): Agroforestry: tenure and incentives. Available from: http://minds.wisconsin.edu/handle/1793/34122.

Bryman A (2006): Integrating quantitative and qualitative research: How is it done? Qualitative Research **6**: 97–113.

Buchanan JM (1958): The Thomas Jefferson Center for studies in political economy. The University of Virginia News Letter **vol. XXXV(2), Oct 15th**: 5–8.

Buchanan JM (1959): Positive Economics, Welfare Economics, and Political Economy. The Journal of Law & Economics **2**: 124–138.

Buchanan JM (1980): Rent seeking and profit seeking. In: Buchanan JM, Tollison RD & Tullock G (eds.) Toward a theory of the rent-seeking society. College Station: Texas A & M University P 3–15.

Buchanan JM (1990): The domain of constitutional economics. Constitutional Political Economy **1**: 1–18.

Buchanan JM & Stubblebine WC (1969): Externality. Economica, New Series **29**: 371–384.

Buchanan JM & Tullock G (1962): The Calculus of Consent. Logical Foundations of Constitutional Democracy. The University of Michigan Press: Ann Arbor.

Burns RB & Burns RA (eds.) (2008): Business research methods and statistics using SPSS. Los Angeles, London: Sage.

Burnside AC & Dollar D (2000): Aid, Policies, and Growth. American Economic Review **90**: 847-868.

Burton TI (2016): Is the Country of Georgia the Next Great Wine Destination? The Wall Street Journal, April 7, 2016, Tbilisi. Available from: https://www.wsj.com/articles/is-the-country-of-georgia-the-next-great-wine-destination-1460045910.

Buschmann A (2008): Headache or risk-aversion: Why peasants in Georgia are reluctant to cooperate formally? Or the role of cooperation in the agriculture of Georgia. (Master Thesis); Javakhishvili Tbilisi State University. Tbilisi.

Business & Finance Consulting (2018): Georgian Agriculture Finance Bulletin, Edition #69, June 2018. Available from: https://georgien.ahk.de/fileadmin/ahk_georgien/Publikation/_BFC-Georgia-Agri-Bulletin-70-JUNE-18.pdf.

Calabresi G (1968): Transaction Costs, Resource Allocation and Liability Rules--A Comment. The Journal of Law and Economics **11**: 67–73.

Calabresi G & Melamed AD (1972): Property rules, liability rules, and inalienability: One view of the Cathedral. Harvard Law Review **85**: 1089–1128.

Campbell DT & Fiske DW (1959): Convergent and Discriminant Validation by the Multitrait-Multimethod Matrix. Psychological Bulletin **56**: 81–105.

Campbell JL (2010): Institutional Reproduction and Change. In: Morgan G, Campbell JL, Crouch C, Perdersen OK & Whitley R (eds.) The Oxford Handbook of Comparative Institutional Analysis. New York: Oxford University Press 87–116.

Cardoso FH & Faletto E (1979): Dependency and development in Latin America. Berkeley: University of California Press.

Carter MR & Olinto P (2003): Getting Institutions "Right" for Whom? Credit Constraints and the Impact of Property Rights on the Quantity and Composition of Investment. American Journal of Agricultural Economics **85**: 173–186.

Cashin SM & McGrath G (2006): Establishing a modern cadastral system within a transition country: Consequences for the Republic of Moldova. Land Use Policy **23**: 629–642.

Cemovich R (2001): Privatization and Registration of Agricultural Land in Georgia. Post-Soviet Affaires **17**: 81–100.

Charmaz K (1996): 3. Grounded Theory. In: Smith, J. A., Harré, R., Van Langenhove (ed.) Rethinking Methods in Psychology. London: Sage 27–49.

Chavleishvili M (2015): Consumer Market for Georgian Hazelnut and the Strategy to Improve Its Competitiveness. World Academy of Science, Engineering and Technology **9**.

Cheung, Steven N. S. (1970): The structure of a contract and the theory of a non-exclusive resource. Journal of Law and Economics **13**: 49–70.

Christophe B (2004a): Metamorphosen des Leviathan in einer post-sozialistischen Gesellschaft, Georgiens Provinz zwischen Fassaden der Anarchie und regulativer Allmacht. Bielefeld: Transcript Verlag.

Christophe B (2004b): Parastaatlichkeit und Schattenglobalisierung: Das Beispiel Georgien. In: Kutenbach S & Lock P (eds.) Kriege als (Über)Lebenswelten, Schattenglobalisierung, Kriegsökonomien und Inseln der Zivilität: EINE Welt, Texte der Stiftung Entwicklung und Frieden. Bonn: Dietz 88–100.

Civil Georgia (2010a): Commissioner Says Georgia Committed to EU Partnership: May 13, 2010. Available from: http://civil.ge/eng/article.php?id=22296&search=ENP.

Civil Georgia (2007a): Government Lays Out Plan for Agro-Business Development: July 25, 2007. Available from: http://www.civil.ge/eng/article.php?id=15494.

Civil Georgia (2007b): Saakashvili Outlines Cabinet's Programme: September 6, 2007. Available from: http://civil.ge/eng/article.php?id=15743.

Coase RH (1960): The Problem of Social Cost. Journal of Law and Economics **Vol. III**: 1–44.

Coase RH (1988): The Firm, the Market, and the Law. In: Coase RH (ed.) The Firm, the Market, and the Law. Chicago: University of Chicago Press 1–30.

Cohen F (1954): Dialogue on Private Property. Available from: http://digitalcommons.law.yale.edu/fss_papers/4360.

Cole DH (2013): The Varieties of Comparative Institutional Analysis. Indiana University Maurer School of Law, Paper 834. Available from: https://papers.ssrn.com/sol3/papers.cfm?abstract_id=2162691.

Cole DH (2017): Laws, norms, and the Institutional Analysis and Development framework. Journal of Institutional Economics **10**: 1–19.

Cole DH & McGinnis MD (eds.) (2017): Elinor Ostrom and the Bloomington School of Political Economy. London: Lexington Books.

Coleman J (1966): The Possibility of a Social Welfare Function. American Economic Review **56**: 1105–1122.

Colson E (1974): Tradition and Contract: The Problem of Order. Chicago: Aldine Publishing Company.

Commons JR ([1924] 1968): Legal Foundations of Capitalism. Madison: University of Wisconsin Press.

Commons JR (1931): Institutional Economics. The American Economic Review **Vol. 21**: 648–657.

Commons JR (1932): The problem of correlating law, economics and ethics. Wisconsin Law Review **8**: 3–26.

Commons JR (1934): Institutional Economics: It's Place in Political Economy. New York: Macmillan.

Cordonnier C (2010): EU Export Market Conditions for the Realisation of the Competitive Advantages of Georgian Agricultural Products . Available from: http://www.geplac.com/newfiles/GeorgianEconomicTrends/Cor-donnier%202010.pdf.

Cornell SE, Swanström N, Tabyshalieva A & Tcheishvili G (2005): A Strategic Conflict Analysis of the South Caucasus with a Focus on Georgia: Report to Swedish Internation Development Cooperation Agency. Washington, Stockholm: Central Asia-Caucasus Institute.

Corso M (2012): Georgia: Can Property Rights Survive the Digital Age? Available from: http://www.eurasianet.org/node/65275.

CSS (2012): The Role of Social Capital in Rural Community Development in Georgia. Centre for Social Studies(CSS)/Academic Swiss Caucasus Net (ASCN), Tbilisi. Available from: http://www.ascn.ch/en/research/Current-Projects/Georgia-2011/mainColumnParagraphs/03/text_files/file/Summary%20Muskhelishvili.pdf.

Dahlman CJ (1979): The Problem of Externality. Journal of Law and Economics **22**: 141–162.

Dahlman CJ (1980): The open-field system and beyond: A propety rights analysis of an economic institution. Cambridge: Cambridge University Press.

Dale PF & Baldwin R (1998): Emerging Land Markets in Central and Eastern Europe. ACE Project P2128R (Action for Co-operation in the field of Economics). Draft Final Report (unpublished).

Dale PF & Baldwin R (2000): Lessons Learnt from the Emerging Land Markets in Central and Eastern Europe: Paper presented at the Quo Vadis - International Conference, FIG Working Week 2000, 21-26 May, Prague. Available from: http://fig.net/resources/proceedings/2000/prague-final-papers/baldwin-dale.htm.

Dale PF & McLaren RA (1999): GIS and land administration. In: Longley PA, Goodchild MF & Maguire, David J., Rhind, David W. (eds.) Geographical information systems. 2nd ed. New York: Wiley 859–875.

Dale PF & McLaughlin JD (1999): Land administration. Oxford, New York: Oxford University Press.

Davis LE & North D (1970): Institutional Change and American Economic Growth: A First Step Towards a Theory of Institutional Innovation. J. Eco. History **30**: 131–149.

Deininger K & Chamorro JS (2004): Investment and equity effects of land regularisation: the case of Nicaragua. Agricultural Economics **30**: 101–116.

Deininger K & Feder G (2001): Land institutions and land markets. In: Gardner B & Rausser G (eds.) Handbook of agricultural economics. Amsterdam: Elsevier 288–331.

Deininger K & Feder G (2009): Land Registration, Governance, and Development: Evidence and Implications for Policy. The World Bank Research Observer **24**: 233–266.

Deininger K, Feder G, Gordillo de Anda G & Munro-Faure P: Land policy to facilitate growth and poverty reductions 5–18.

Demsetz H (1967): Toward a theory of property rights. American Economic Review **57**: 347–359.

Denzin NK (1970): The Research Act in Sociology. A Theoretical Introduction to Sociological Methods. Chicago: Aldine.

Denzin NK (1978): The Research Act: A Theoretical Introduction to Sociological Methods. 2nd. New York: McGraw Hill.

Dey I (1993): Qualitative data analysis: A user-friendly guide for social scientists. London, New York: Routledge.

Di Gregorio M, Hagedorn K, Kirk M, Korf B, McCarthy N, Meinzen-Dick R & Swallow B (2008): Property rights, collective action and poverty: the role of institutions for poverty reduction. Available from: http://www.capri.cgiar.org/pdf/capriwp81.pdf.

Di John J (2008): Conceptualising the Causes and Consequences of Failed States: A Critical Review of the Literature. Working Paper No. 25, Crisis States Research Centre, London School of Economics, London. Available from: https://www.files.ethz.ch/isn/57427/wp25.2.pdf.

Didebulidze A & Plachter H (2002): Nature conservation aspects of pastoral farming in Georgia. In: Redecker B, Finck P, Härdtle W, Riecken U & Schröder, E (eds.) Pasture landscapes and nature conservation. Berlin: Springer 87–105.

Didebulidze A & Urushadze T (2009): Agriculture and land use change in Georgia. In: King L & Khubua G (eds.) Georgia in Transition: Experiences and Perspectives. Schriften zur internationalen Entwicklungs- und Umweltforschung. Zentrum für internationale Entwicklungs- und Umweltforschung der Justus-Liebig-Universität Gießen 241–263.

Dirimanova V (2005): Land Market with Fragmented Landownership Rights in Bulgaria: An Institutional Approach: Paper presented at the 94th EAAE Seminar 'From households to firms with independent legal status: the spectrum of institutional units in the development of European agriculture ', Ashford (UK), 9-10 April 2005.

Djelic M-L & Quack S (2007): Overcoming path dependency: Path generation in open systems. Theor Soc **36**: 161–186.

Douglas M (1973): Rules and meanings: The anthropology of everyday knowledge. Penguin modern sociology readings. [Harmondsworth, Eng.]: Penguin Books.

Douglas M (1986): How institutions think. 1st ed. The Frank W. Abrams lectures. Syracuse, N.Y.: Syracuse University Press.

Downs A (1957): An Economic Theory of Political Action in a Democracy. Journal of Political Economy **65**: 135–150.

Dowson E & Sheppard VLO (1952): Land registration. London: H. M. Stationery Off.

DWVG (2013a): DWVG Agro & Food News 07/02/13: Farmers Will Get Agricultural Vouchers Of Different Value.
DWVG (2013b): Newsletter Georgien, Februar 2013: 700 Mio. GEL sollen dem Agriculture Development Fund zu Verfügung stehen.
DWVG (2011d): DWVG Agro & Food News Georgia 01/11/2011: Agriculture Of Georgia Is Mainly Funded By Donors.
DWVG (2011c): Agro & Food News in Georgia 2011, 5/08/11: Boers bought 98 hectares of land.
DWVG (2011b): DWVG Agro & Food News 16/02/11: GEL 150 mln Extra Funding Planned for Agriculture
DWVG (2011a): DWVG Agro & Food News Georgia 02/02/11: Kviris Palitra: "Peasantry Rejects The Land".
Easterly W (2003): Can Foreign Aid Buy Growth? Journal of Economic Perspectives **17**: 23–48.
Ebanoidze J (2003): Current Land Policy Issues in Georgia. In: FAO & World Bank (eds.) Land reform - land settlement and cooperatives - Special Edition 125–140.
Ebbinghaus B (2005): Can Path Dependence Explain Instutional Change? Two Approaches Applied to Welfare State Reform. MPIfG Discussion Paper 05/2, Cologne, Germany. Available from: www.mpifg.de/pu/mpifg_dp/DP05-2.pdf.
EBRD (2006): Strategy for Georgia. Available from: http://www.ebrd.com/about/strategy/country/georgia/strategy.pdf.
Edilashvili N (2012): USAID launches nuts training program. Georgia Today, Issue 599. Available from: http://old.georgiatoday.ge/article_details.php?id=9872.
Eggertsson T (1990): Economic behavior and institutions. Cambridge: Cambridge University Press.
Egiashvili D (2002): Strategy for Land Consol Consolidation and Improved Land Management in Georgia. State Department of Land Management of Georgia. Available from: www.fao.org/fileadmin/user_upload/Europe/…2002/Land2002/georgia_paper.pdf.
Egiashvili D (2012): LGAF Georgia: The Best Practice - Property Registry Reform. Available from: http://siteresources.worldbank.org/INTLGA/Resources/LGAF_Georgia_BestPractice_PropertyRegistryReform_Oct2012.pdf.
Egiashvili D (2013): Issues and Options for Improved Land Sector Governance in Georgia: Application of the Land Governance Assessment Framework. Synthesis Report.
Ellickson RC (1986): Of Coase and Cattle: Dispute Resolution Among Neighbors in Shasta County. Stanford Law Review **38**: 623–687.
Enemark S (2009): Land Administration Systems - managing rights, restrictions and responsibilities in land: Presented at the Map World Forum, Hyderabad, India, 10-13 February 2009. Available from: https://www.fig.net/organisation/council/council_2007-2010/council_members/enemark_papers/2009/hyderabad_enemark_paper_feb_2009.pdf.
FAO (2007): Good governance in land tenure and administration. FAO land tenure studies, 9. Rome: Food and Agriculture Organization of the United Nations.
Farr J (2004): Social Capital. Political Theory **32**: 6–33.
Feder G & Onchan T (1987): Land Ownership Security and Farm Investment in Thailand. American Journal of Agricultural Economics **69**: 311–320.
Field E (2005): Property Rights and Investment in Urban Slums. Journal of the European Economic Association **3**: 279–290.
Fielding NG & Fielding JL (1986): Linking Data: Qualitative Re-search Methods. Beverly Hills et al.: Sage.

FIG (2010): Sydney Declaration: Declaration from the XXIV FIG Congress, Sydney, Australia, 11-16 April 2010. Available from: http://fig.net/resources/publications/figpub/sydney_decl/sydney_declaration.pdf.

Firmin-Sellers K & Sellers P (1999): Expected Failures and Unexpected Successes of Land Titling in Africa. World Development **27**: 1115–1128.

Flick U (1992): Triangulation Revisited: Strategy of Validation or Alternative? J Theory of Social Behaviour **22**: 175–197.

Flick U (1995): Triangulation. In: Flick U, Kardorff Ev, Keupp H, Rosenstiel L von & Wolff S (eds.) Handbuch qualitative Sozialforschung: Grundlagen, Konzepte, Methoden und Anwendungen. 2. Aufl. Grundlagen Psychologie. Weinheim: Beltz 432–434.

Freedom House (2005): Georgia: Nations in Transit. Available from: https://freedomhouse.org/report/nations-transit/2005/georgia#.VdHhjpeGydE.

Furubotn E & Pejovich S (1972): Property rights and economic theory: A survey of recent literature. Journal of Economic Literature **10**: 1137–1162.

Galiani S & Schargrodsky E (2010): Property rights for the poor: Effects of land titling. Journal of Public Economics **94**: 700–729.

Gegeshidze A (2012): Georgia's Political Transformation in Zigzag. In: Friedrich-Ebert-Stiftung (ed.) South Caucasus - 20 Years of Independence. Berlin: Friedrich-Ebert-Stiftung.

Georgian National Museum (2014): Supra - Feiern auf Georgisch. Georgische Kulturtage im Museum Europäischer Kulturen, 1.8. - 30.8.2014, Berlin. Available from: http://www.museumsportal-berlin.de/en/events/supra-feiern-auf-georgisch/slideshow/#0.

Georgien Nachrichten (2011): Erst Feierstunde, dann Enteignung: 270 Eigentümer verlieren Land in Georgien: November 24, 2011. Available from: http://www.georgien-nachrichten.de/index.php?rubrik=panorama&cmd=n_einzeln&nach_id=18067&anfang=256.

Georgien Nachrichten (2010): Verkauf von Grundstücken in Georgien: Welle von Gewalt und Verhaftungen in Mestia: July 12, 2010. Available from: http://www.georgien-nachrichten.de/index.php?rubrik=innenpolitik&cmd=n_einzeln&nach_id=17688.

GeoStat (2017): Agriculture. National Statistics Office of Georgia, Tbilisi. Available from: http://geostat.ge/index.php?action=page&p_id=428&lang=eng.

Gerring J (2004): What Is a Case Study and What Is It Good for? The American Political Science Review **98**: 341–354.

Ghatak M & Roy S (2007): Land reform and agricultural productivity in India: a review of the evidence. Oxford Review of Economic Policy **23**: 251–269.

Gittell R & Vidal A (1998): Community Organizing: Building social capital as a development strategy. Thousand Oaks, California: Sage Publications.

Glaser BG & Strauss AL (1965): Awareness of dying. Chicago: Aldine.

Glaser BG & Strauss AL (1967a): The discovery of grounded theory: Strategies for qualitative research. New York, N.Y.: Aldine Publishing Co.

Glaser BG & Strauss AL (1967b): The Discovery of Grounded Theory. Mill Valley: The Sociology Press.

Glaser BG & Strauss AL (1968): Time for dying. Chicago: Aldine Publishing.

Gomulka S (2007): Economic and political constraints during transition. Europe-Asia Studies **46**: 89–106.

Gore C (2000): The Rise and Fall of the Washington Consensus as a Paradigm for Developing Countries. World Development **28**: 789–804.

Granovetter M (1985): Economic Action and Social Structure: The Problem of Embeddedness. American Journal of Sociology **91**: 481–510.

GSH (2017): Samegrelo. Georgiastartshere.com. Available from: http://georgiastartshere.com/samegrelo/.

GTAI (2015): Wirtschaftsstruktur und -chancen: Georgien. Available from: https://www.gtai.de/GTAI/Content/DE/Trade/Fachdaten/PUB/2015/03/pub201503188001_19812_wirtschaftsstruktur-und--chancen---georgien--2015.pdf.

GTZ/CIPDD (2006): Potential for Conflict Related to Land Problems in Georgia's Marneuli and Gardabani Districts. Available from: http://www.cipdd.org/files/45_171_803701_LandProblem-Eng%5B1%5D.pdf.

Gunder Frank A (1966): The Development of Underdevelopment. Mon. Rev. **18**: 17.

Gvaramia A (2013): Land Ownership and the Development of the Land Market in Georgia: A Report Commissioned by Alliances KK and Undertaken by a Private Consultant. Available from: https://www.eda.admin.ch/content/dam/countries/countries-content/georgia/en/resource_en_219898.pdf.

GWA (2016): Grape harvest 2016. Georgian Wine Association. Available from: http://www.gwa.ge/?3/554/.

Gylfason T & Hochreiter E (2009): Growing Apart? A Tale of Two Republics: Estonia and Georgia. European Journal of Political Economy **25**: 355-270.

Hage W (2007): Das orientalische Christentum. Religionen der Menschheit, Bd. 29, 2. Stuttgart: Verlag W. Kohlhammer.

Hagedorn K (1991): Public Choice and Agricultural Policy. In: Sen A (ed.) Issues in contemporary economics proceedings of the Ninth World Congress of the International Economic Association, Issues in Contemporary Economics, Proceedings of the Ninth' World Congress of the International Economic Association. Athens, Greece.

Hagedorn K (1992): Wirtschaftliche und politische Treibkräfte der Umformung einer sozialistischen Agrarverfassung. In: Boettcher E, Herder-Dorneich P, Schenk K-E & Schmidtchen D (eds.) Jahrbuch für Neue Politische Ökonomie. Bd. 11. Tübingen: J.C.B. Mohr (Paul Siebeck) 191–213.

Hagedorn K (1994): Interest Groups. In: Hodgson G, Tool M & Samuels WJ (eds.) Handbook of Institutional and Evolutionary Economics. Cheltenham: Edward Elgar.

Hagedorn K (1996): Das Institutionenproblem in der agrarökonomischen Politikforschung. Schriften zur Angewandten Wirtschaftsforschung, Bd. 72. Tübingen: J.C.B. Mohr (Paul Siebeck).

Hagedorn K (2004): Property rights reform on agricultural land in Central and Eastern Europe. Quarterly Journal of International Agriculture **43**: 409–438.

Hagedorn K (2005): Integrative and Segregative Institutions: a dichotomy for understanding institutions of sustainability: Paper to be presented at the Workshop in Political Theory and Policy Analysis, Colloquium Mini Series, Bloomington, October 20, 2005. Available from: http://www.indiana.edu/~workshop/colloquia/materials/fall2005_colloquia.html#hagedorn.

Hagedorn K (2008): Particular requirements for institutional analysis in nature-related sectors. European Review of Agricultural Economics **35**: 357–384.

Hagedorn K & Arzt, K., Peters, U. (2002): Institutional arrangements for environmental co-operatives: a conceptual framework. In: Hagedorn K (ed.) Environmental Co-operation and Institutional Change: Theories and Policies for European Agriculture. Cheltenham, UK: Edward Elgar.

Hagedorn K & Schmitt G (1985): Die politischen Gründe für eine wirtschaftspolitische Vorzugsbehandlung der Landwirtschaft. In: Boettcher E, Herder-Dorneich P & Schenk K-E (eds.) Jahrbuch für Neue Politische Ökonomie. Bd. 4. Tübingen: J.C.B. Mohr (Paul Siebeck) 250–295.

Halbach U (2006): Säbelrasseln und Friedenspolitik in Europas neuer Nachbarschaft: SWP-Aktuell 32. Available from: http://www.swp-berlin.org/de/publikationen/swp-aktuell-

de/swp-aktuell-detail/article/saebelrasseln_und_friedenspolitik_in_europas_neuer_nachbarschaft.html.
Halbach U (2007): Die Krise in Georgien: Das Ende der „Rosenrevolution“?: SWP-Aktuell 2007/A 61.
Halbach U (2008): Politik im Südkaukasus. Available from: http://www.swp-berlin.org/fileadmin/contents/products/aktuell/2008A31_hlb_ks.pdf.
Halbach U & Müller F (2001): Persischer Golf, Kaspisches Meer und Kaukasus: Entsteht eine Region strategischen europäischen Interesses?
Hansen H & Tarp F (2000): Aid effectiveness disputed. Journal of International Development **12**: 375–398.
Harbison JS (1992): Hohfeld and Herefords: The Concept of Property and the Law of the Range. New Mexico Law Review **22**: 459–499.
Hardin G (1968): The tragedy of the commons. Science **162**: 1243–1248.
Harriss-White B & Janakarājan S (eds.) (2004): Rural India facing the 21st century. Anthem South Asian studies. London: Anthem Press.
Harvey DI, Kellard NM, Madsen JB & Wohar ME (2010): The Prebisch-Singer Hypothesis: Four Centuries of Evidence. Review of Economics and Statistics **92**: 367–377.
Haydu J (1998): Making Use of the Past: Time Periods as Cases to Compare and as Sequences of Problem Solving. American Journal of Sociology **104**: 339–371.
Heynen N, McCarthy J, Prudham S & Robbins P (eds.) (2007): Neoliberal environments: False promises and unnatural consequences. London, New York: Routledge.
Ho P & Spoor M (2006): Whose Land? The Political Economy of Land Titling in Transitional Economies. Land Use Policy **23**: 580–587.
Hodgson G (2006): What are Institutions? Journal of Economic Issues **XL**: 1–25.
Hoebel EA (1942): Fundamental Legal Concepts as Applied in the Study of Primitive Law. Yale Law Journal **51**: 951–956.
Hohfeld WN (1913): Some Fundamental Legal Conceptions as Applied in Judicial Reasoning. Yale Law Journal **23**: 16–59.
Hohfeld WN (1917): Fundamental Legal Conceptions as Applied in Judicial Reasoning. Yale Law Journal **26**: 710–770.
Honoré AM (1961): Ownership. In: Guest AG (ed.) Oxford Essays in Jurisprudence. Oxford: Oxford University Press Chapter 5.
ICG (2007): Georgia: Sliding towards authoritarianism? Europe Report N°189. Available from: http://www.crisisgroup.org/home/index.cfm?id=5233, downloaded on December 22, 2007.
IFAD (2007): Republic of Georgia, Agricultural Development Project Completion Evaluation: Report N° 1853-GE. Available from: http://www.ifad.org/evaluation/public_html/eksyst/doc/prj/region/pn/georgia/georgia.pdf.
Ikdahl I, Hellum A, Kaarhus R & Benjaminson Tor A (2005): Human rights, formalisation and women's land rights in southern and eastern Africa. Noragric Report No. 26, Noragric, Norwegian University of Life Sciences. Available from: http://www.umb.no/statisk/noragric/publications/reports/2005_nor_rep_26.pdf.
IUCN (2012): Georgia – a haven for biodiversity. International Union for Conservation of Nature, 7.12.2012. Available from: https://www.iucn.org/content/georgia-%E2%80%93-haven-biodiversity.
Jakob A (2001): On the Triangulation of Quantitative and Qualitative Data in Typological Social Research: Reflections on a Typology of Conceptualizing "Uncertainty" in the Context of Employment Biographies. Forum Qualitative Sozialforschung/Forum: Qualitative Social Research, Vol. 2, No. 1. Available from: http://www.qualitative-research.net/index.php/fqs/article/view/981.

Janvry A de & Sadoulet E (2001): Access to land and land policy reforms. Available from: https://www.staff.ncl.ac.uk/david.harvey/AEF806/AccessToLand.pdf.

Jobelius M (2012): Georgia's authoritarian liberalism. In: Friedrich-Ebert-Stiftung (ed.) South Caucasus - 20 Years of Independence. Berlin: Friedrich-Ebert-Stiftung 77–91.

Jones SF (2000): Democracy from Below? Interest Groups in Georgian Society. Slavic Review **Vol. 59**: 42–73.

Kahneman D & Tversky A (1979): Prospect Theory: An Analysis of Decision under Risk. Econometrica **47**: 263–291.

Kajrishvili T (2014): Rtveli – Harvesting grapes, Georgian style. Geogian Journal, October 16, 2014. Available from: https://www.georgianjournal.ge/discover-georgia/28511-rtveli-harvesting-grapes-georgian-style.html.

Kan I, Kimhi A & Lerman Z (2006): Farm output, non-farm income and commercialization in rural Georgia. Available from: ftp://ftp.fao.org/docrep/fao/009/ah758e/ah758e00.pdf.

Katz EG (2000): Social Capital and Natural Capital: A Comparative Analysis of Land Tenure and Natural Resource Management in Guatemala. Land Economics **76**: 114.

Kaufmann R (2003): Georgien: Ein Reise-Lesebuch. Bruchsal: ERKA-Verl.

Kaufmann R (2015): Zwischen Regierungslyrik und Wirklichkeit. Kaukasische Post, 20. Oktober 2015, Tbilisi. Available from: http://www.kaukasische-post.com/?p=1941.

Kautsky K (1921): Georgia: A Social-Democratic Peasant Republic - Impressions And Observations. London, Bradford, Manchester: International Bookshops Limited.

Kegel H (2003): The Significance of Subsistence Farming in Georgia as an Economic and Social Buffer. In: Abele S & Frohberg K (eds.) Subsistence Agriculture in Central and Eastern Europe: How to Break the Vicious Circle? Studies on the Agricultural and Food Sector in Central and Eastern Europe, Vol. 22 147–160.

Keys D (2003): Now that's what you call a real vintage: professor unearths 8,000-year-old wine. The Independent, 28.12.2003. Available from: http://accuca.conectia.es/ind281203.htm.

KfW (2011): Georgia: Cadastre and Land Register Phases I and II. Available from: http://www.kfw-entwicklungsbank.de/ebank/EN_Home/Evaluation/Ex-post_evaluation_reports/PDF-Dokumente_E-K/Georgien_Kataster_Grundbuch_2011_E.pdf.

Khan M (1995): State Failure in Weak State: A Critique of New Institutionalist Explanations. In: Harriss J, Hunter J & Lewis CM (eds.) The New Institutional Economics and Third World Development. London: Routledge 71–86.

Knight AW (1993): Beria: Stalin's first lieutenant. Princeton, N.J.: Princeton University Press.

Krikorian O (2011): Georgia: Return of the Meskhetian Turks. Available from: http://globalvoicesonline.org/2012/01/03/georgia-return-of-the-meskhetian-turks/.

Krueger A (1974): The political economy of the rent-seeking society. American Economic Review **64**: 291–303.

Krueger A (1988): The Political Economy of Controls: American Sugar. Cambridge, MA: National Bureau of Economic Research.

Krugman P (2008): Inequality and Redistribution. In: Serra N & Stiglitz JE (eds.) The Washington Consensus reconsidered: Towards a new global governance. The initiative for policy dialogue series. Oxford, New York: Oxford University Press 31–40.

Kuhnen F (1982): Agrarverfassungen. In: Blanckenburg Pv (ed.) Sozialokonomie der landlichen Entwicklung. 2nd ed. Stuttgart: Ulmer 69–85.

Kurdadze T (2015): Hazelnut - Georgia's Precious Commodity. Georgia Today, October 8, 2015, Tbilisi. Available from: http://georgiatoday.ge/uploads/images/Figures_1-2_10.9.15_4181.jpg.

Larsson G (1991): Land registration and cadastral systems: Tools for land information and management. Harlow, Essex, England, New York: Longman Scientific and Technical; Wiley.
Lasswell HD (1971): A pre-view of policy sciences. New York: American Elsevier.
Laux G (2005): Landmanagement in Georgien - ein Projekt der GTZ. In: Schröder B (ed.) Georgien: Gesellschaft und Religion an der Schwelle Europas. Saarbrücken: Röhrig 191-.
Lerman Z (1996): Land Reform and Private Farms in Georgia: 1996 Status, EC4NR Agriculture Policy Note No. 6, The World Bank.
Lerman Z (1999): Land Reform and Farm Restructuring: What Has Been Accomplished to Date? American Economic Review **Vol. 89**.
Lerman Z (2004a): Policies and institutions for commercialization of subsistence farms in transition countries. Journal of Asian Economics **15**: 461–479.
Lerman Z (2004b): Successful land individualization in Trans-Caucasia: Armenia, Azerbaijan, Georgia. In: Macey DAJ, Pyle W & Wegren SK (eds.) Building Market Institutions in Post-Communist Agriculture: Land, Credit, and Assistance. Oxford: Lexington Books 53–75.
Lerman Z (2005): Farm Fragmentation and Productivity: Evidence from Georgia. Available from: http://departments.agri.huji.ac.il/economics/en/publications/discussion_papers/2005/lerman-gru.pdf.
Lewis WA (1954): Economic Development with Unlimited Supplies of Labor. Manchester School of Economic and Social Studies **22**: 139–191.
Lipton M (1977): Why Poor People Stay Poor: Urban Bias in World Development. The Scandinavian Journal of Economics **80**.
Llewellyn KN (1960): The Bramble Bush: The Classic Lectures on the Law and Law School. New York, N.Y.: Oceana Publications.
Loehr D (2012): Capitalization by formalization? – Challenging the current paradigm of land reforms. Land Use Policy **29**: 837–845.
Lohm H (2006): Dukhobors in Georgia: A Study of the Issue of Land Ownership and Inter-Ethnic Relations in Ninotsminda rayon (Samtskhe-Javakheti).
Low B, Costanza R, Ostrom E, Wilson J & Simon CP (1999): Human–ecosystem interactions: a dynamic integrated model. Ecological Economics **31**: 227–242.
Lund C (2006): Twilight Institutions: Public Authority and Local Politics in Africa. Development & Change **37**: 685–705.
Lutsevych O (2013): How to Finish a Revolution: Civil Society and Democracy in Georgia, Moldova and Ukraine. Briefing Paper REP BP 2013/01, Chatham House. Available from: http://www.chathamhouse.org/sites/default/files/public/Research/Russia%20and%20Eurasia/0113bp_lutsevych.pdf.
Macours K & Swinnen JFM (2000): Causes of Output Decline in Economic Transition: The Case of Central and Eastern European Agriculture. Journal of Comparative Economics **28**: 172–206.
Mahoney J (2000): Path Dependence in Historical Sociology. Theor Soc **29**: 507–548.
Mahoney J & Thelen KA (2010): Explaining institutional change: Ambiguity, agency, and power. Cambridge, New York: Cambridge University Press.
Mamardashvili P (2008): Analyse der Wettbewerbsfähigkeit georgischer Agrarprodukte am Beispiel von Wein. Masterarbeit, Swiss Federal Institute of Technology Zurich ETH, October 2008, Zurich.
Mardin S (1969): Power, Civil Society and Culture in the Ottoman Empire. Comparative Studies in Society and History **11**: 258–281.

Marengo, L., Pasquali, C. and Valente, M. (2001): Decomposability and modularity of economic interactions: Paper for the Workshop 'Modularity: Understanding the Development and Evolution of Complex Natural Systems', Altenberg, Austria, October 26–29, 2000.
Markussen T (2008): Property Rights, Productivity, and Common Property Resources: Insights from Rural Cambodia. World Development **36**: 2277–2296.
Marr (1976): Early Processing of Visual Information. Philosophical Transactions of the Royal Society of London. Series B, Biological Sciences **275**: 483–519.
McGovern PE (2007): Ancient wine: The search for the origins of viniculture. 4th print., and 1st apperback print. Princeton: University Press.
McHugh ML (2013): The Chi-square test of independence. Biochemia Medica **23**: 143–149.
Mead GH (1934): Mind, Self and Society: From the standpoint of a social behaviorist. Chicago, Ill.: University of Chicago Press.
Méda D (2002): Le capital social: un point de vue critique. L' Economie Politique **2**: 36–47.
Ménard C (2004a): The Economics of Hybrid Organizations. Journal of Institutional and Theoretical Economics **160**: 345–376.
Ménard C (2004b): The Economics of Hybrid Organizations. Journal of Institutional and Theoretical Economics **160**: 345–376.
Migot-Adholla SE, Hazell P, Blarel B & Place F. (1991): Indigenous Land Rights Systems in Sub-Saharan Africa: A Constraint on Productivity? The World Bank Economic Review **5**: 155–175.
Mishan EJ (1974): The economics of disamenity. Natural Resources Journal **14**: 55–86.
Mishkin F (1999): International Capital Movements, Financial Volatility and Financial Instability. NBER Working Paper Series, 6390. Cambridge, MA: National Bureau of Economic Research.
Mitchell LA (2009): Compromising democracy: State building in Saakashvili's Georgia. Central Asian Survey **28**: 171–183.
Mitchell WC (1988): Virginia, Rochester, and Bloomington: Twenty-five years of public choice and political science. Public Choice **56**: 101–119.
MoA (2015): Annual Report 2015. Ministry of Agriculture of Georgia. Available from: http://www.moa.gov.ge/Download/Files/187.
MoA (2016): Technical regulations on hazelnut products have been approved at the Georgian Government session. Ministry of Agriculture, April 14, 2016. Available from: http://moa.gov.ge/En/News/1090.
Mohanty CT (1991): Under Western Eyes. Feminist Scholarship and Colonial Discourse. In: Mohanty CT, Russo A & Torres L (eds.) Third world women and the politics of feminism. Bloomington: Indiana Univ. Press.
Moore SF (1978): Law as process: An anthropological approach. London, Boston: Routledge & K. Paul.
Mühlfried F (2010): Did communism matter? Settlement policies from above and below in Highland Georgia. In: Tsitsishvili N (ed.) Cultural Paradigms and Political Change in the Caucasus. Saarbrücken: Lambert 174–196.
Narmania D (2009): Economic policy in Georgia: Liberalization, Economic Crisis and Changes.
Navia P & Velasco A (2003): The Politics of Second-Generation Reforms. In: Kuczynski Godard P-P & Williamson J (eds.) After the Washington consensus: Restarting growth and reform in Latin America. Washington, DC: Institute for International Economics 265–303.
Nelson RR & Winter SG (1982): An evolutionary theory of economic change. Cambridge, Massachusetts, London: Belknap Press of Harvard University Press.

Nichols S (1993): Land registration: managing information for land administration: Ph.D. dissertation, Department of Surveying Engineering Technical Report No. 168, University of New Brunswick, Fredericton, New Brunswick, Canada.
Nielsen F (2000): Wind, der weht. Georgien im Wandel. Frankfurt: Societäts-Verlag.
Nielsen K, Jessop B & Hausner J (1995): Institutional change in post-socialism. In: Hausner J, Jessop B & Nielsen K (eds.) Strategic choice and path-dependency in post-socialism: Institutional dynamics in the transformation process. Aldershot, UK, Brookfield, Vt., US: E. Elgar 1–9.
Niskanen WA (1971): Bureaucracy and Representative Government. Chicago: Aldine-Atherton.
Noor KBM (2008): Case Study: A Strategic Research Methodology. American Journal of Applied Sciences **5**: 1602–1604.
North D (1993): Five propositions about institutional change: Essay presented at the Economic History Workshop at Stanford University. Available from: http://econwpa.repec.org/eps/eh/papers/9309/9309001.pdf.
North DC (1990a): A Transaction Cost Theory of Politics. Journal of Theoretical Politics **2**: 355–367.
North DC (1990b): Institutions, institutional change, and economic performance. The Political economy of institutions and decisions. Cambridge, New York: Cambridge University Press.
North DC (1991): Institutions. Journal of Economic Perspectives **5**: 97–112.
North DC & Thomas RP (1977): The First Economic Revolution. Economic History Rev **30**: 229–241.
Olsen W (2004): Triangulation in Social Research: Qualitative and Quantitative Methods Can Really Be Mixed. Forthcoming as a chapter in: M. Holborn (ed.) (2004): Developments in Sociology, Ormskirk: Causeway Press.
Olson M (1965): The Logic of Collective Action: Public Goods and the Theory of Groups. Cambridge, Massachusetts: Harvard University Press.
Ortoidze T & Schumann F (2014): Deutsche Siedler als Weinbauern!
Ostrom E (2005): Understanding institutional diversity. Princeton: Princeton University Press.
Ostrom E (2007): Institutional rational choice: An assessment of the Institutional Analysis and Development Framework. In: Sabatier PA (ed.) Theories of the policy process. 2nd ed. Boulder (CO): Westview Press 21–64.
Ostrom E (2008): Governing the Commons: The Evolution of Institutions for Collective Action. 22nd. (Political Economy of Institutions and Decisions). Cambridge: Cambridge University Press.
Ostrom E (2011): Background on the Institutional Analysis and Development Framework. Policy Studies Journal **39**: 7–27.
Ostrom E, Gardner R, Walker J & Agrawal A (1994): Rules, games, and common-pool resources. Ann Arbor: University of Michigan Press.
Ostrom E, Gibson C, Shivakumar S & Andersson K (2014): An Institutional Analysis of Development Cooperation. In: Barrett S, Mäler K-G & Maskin ES (eds.) Environment and Development Economics. Essays in honour of Partha Dasgupta. Oxford: Oxford University Press 117–141.
Ostrom V (1975): Competition, Monopoly, and the Organization of Government in Metropolitan Areas: Comment. Journal of Law and Economics **18**: 691–694.
Ostrom V (1976): The American experiment in constitutional choice. Public Choice **27**: 1–12.
Ostrom V & Ostrom E (1971): Public Choice: A Different Approach to the Study of Public Administration. Public Administration Review **31**: 203–216.

Ostrom V, Tiebout CM & Warren R (1961): The organization of government in metropolitan areas: a theoretical inquiry. Am Polit Sci Rev **55**: 831–842.

Ouedraogó S, Sawadogo J-P, Stamm V & Thiombiano T (1996): Tenure, agricultural practices and land productivity in Burkina Faso: some recent empirical results. Land Use Policy **13**.

Pallot J (2005): Russia's penal peripheries: space, place and penalty in Soviet and post-Soviet Russia. Transactions of the Institute of British Geographers **30**: 98–112.

Papava V (2003): On the Role of the International Monetary Fund in the Post-Communist Transformation of Georgia. Emerging Markets, Finance, and Trade **Vol. 39**: 5–26.

Peters I, Christoplos I, Funder M, Esbern F-H & Pain A (2012): Understanding institutional change: A Review of Selected Literature for the Climate Change and Rural Institutions Research Programme. DIIS Working Paper 2012: 12. Available from: https://www.files.ethz.ch/isn/154356/WP2012-12-%20literature-review-CCRI_mfu_ich_web.pdf.

Petracco CK & Pender J (2009): Evaluating the impact of land tenure and titling on access to credit in Uganda. Available from: http://www.ifpri.org/publication/evaluating-impact-land-tenure-and-titling-access-credit-uganda.

Pigou AC (1920): The Economics of Welfare. London: Macmillan.

Pinckney TC & Kimuyu PK (1994): Land Tenure Reform in East Africa: Good, Bad, or Unimportant. Journal of African Economies **3**: 1–28.

Place F & Hazell P (1993): Productivity Effects of Indigenous Land Tenure Systems in Sub-Saharan Africa. American Journal of Agricultural Economics **75**: 10–19.

Place F & Migot-Adholla SE (1989): The Economic Effects of Land Registration on Smallholder Farms in Kenya: Evidence from Nyeri and Kakamega Districts. Land Economics **74**: 360–373.

Platteau J-P (1996): The Evolutionary Theory of Land Rights as Applied to Sub-Saharan Africa: A Critical Assessment. Development & Change **27**: 29–86.

Polanyi M (1967): The Tacit Dimension. Garden City (NY): Anchor Books.

Polski MM & Ostrom E (1999): An institusional framework for policy analysis and design. Workshop in Political Theory and Policy Analysis, Indiana University. Available from: https://mason.gmu.edu/~mpolski/documents/PolskiOstromIAD.pdf.

Posner RA (1975): The Social Costs of Monopoly and Regulation. Journal of Political Economy **83**: 807–827.

Posner RA (1977): Economic Analysis of Law. Boston: Little Brown.

Prasad C (2012): South African Boers in Georgia?

Prein G, Kelle U & Kluge S (1993): Strategien zur Integration quantitativer und qualitativer Auswertungsverfahren. Arbeitspapier 19 des SFB 186, Universität Bremen. Available from: http://www.sfb186.uni-bremen.de/download/paper19.pdf.

Radelet S (2006): A primer on foreign aid, Working Paper No. 92, Center for Global Development: Washington. Available from: https://www.cgdev.org/sites/default/files/8846_file_WP92.pdf.

Ramstad Y (1990): The Institutionalism of John R. Commons: Theoretical Foundations of a Volitional Economics. In: Samuels WJ (ed.) Research in the History of Economic Thought and Methodology. Boston: JAI Press 53–106.

Reichertz J (2009): Abduction: The Logic of Discovery of Grounded Theory. Forum: Qualitative Social Research **11**: 1–11.

Reichertz J (2010): Abduction: The Logic of Discovery of Grounded Theory. Forum: Qualitative Social Research, Vol. 11, No. 1, Art. 13, January 2010. Available from: http://www.qualitative-research.net/index.php/fqs/article/view/1412/2902.

Reisner O & Kvatchadze L (2005): Georgien. Available from: http://library.fes.de/pdf-files/id/04432.pdf.
Revishvili Z & Kinnucan HW (2004): Agricultural Problems in Georgia and Strategic Policy Responses. In: Petrick M/WP (ed.) The Role of Agriculture in Central and Eastern European Rural Development: Engine of Change or Social Buffer?: IAMO 52–64.
Riker WH (1962): The theory of political coalitions. New Haven: Yale University Press.
Rodrik D (2002): After Neoliberalism, What?: Remarks at the BNDES Seminar on "New Paths of Development", Rio de Janeiro, September 12-13. 2002.
Rodrik D (2004): Institutions and Economic Performance - Getting Institutions Right: CESifo DICE Report 2/2004.
Roland G (2000): Transition and economics: Politics, markets, and firms. Comparative institutional analysis. Cambridge, MA: MIT Press.
Rolfes, Jr., Leonard & Grout M (2013): Assessment of citizens' property rights in Georgia. A report produced for Landesa. Available from: http://ewmi-prolog.org/images/files/4480ENG_-_Assessment_of_Property_Rights_in_Georgia.pdf.
Rostow WW (1971): The Stages of Economic Growth: A Non-Communist Manifesto. 2d ed. Cambridge: University Press.
Rubin A & Babbie ER (2010): Essential research methods for social work. 2nd ed. Belmont, CA: Cengage Learning.
Ruttan VW & Hayami Y (1984): Toward a theory of induced institutional innovation. The Journal of Development Studies **20**: 203–223.
Salukvadze J (2006): Increasing Role of Geoinformation Technologies in Land Management and Beyond: Case of Georgia: Shaping the Change, XXIII FIG Congress, Munich, Germany, October 8-13, 2006. Available from: http://www.fig.net/pub/fig2006/papers/plen03/plen03_02_salukvadze_0708.pdf.
Samuels WJ (1971): Interrelations between Legal and Economic Processes. Journal of Law and Economics **14**: 435–450.
Samuels WJ (1972): In Defense of a Positive Approach to Government as an Economic Variable. Journal of Law and Economics **15**: 453–459.
Saunders B, Kitzinger J & Kitzinger C (2015): Anonymising interview data: challenges and compromise in practice. Qualitative Research **15**: 616–632.
Schmid AA (2004): Conflict and Cooperation: Institutional and Behavioral Economics. Oxford: Wiley-Blackwell.
Schweigert T (2006): Land Title, Tenure Security, Investment and Farm Output: Evidence from Guatemala. The Journal of Developing Areas **40**: 115–126.
Sehring J (2009): Path Dependencies and Institutional Bricolage in Post-Soviet Water Governance. Water Alternatives **2**: 61–81.
Shepsle KA (1989): Studying Institutions. Journal of Theoretical Politics **1**: 131–147.
Shepsle KA & Weingast BR (1984): When Do Rules of Procedure Matter? The Journal of Politics **46**: 206–221.
Shubik M (1982): Game Theory in the Social Sciences. Cambridge, Mass.: MIT Press.
Shubik M (1986): The Games within the Game: Modeling Politico-Economic Structures. In: Kaufmann F-X, Majone G & Ostrom V (eds.) Guidance, control, and evaluation in the public sector: The Bielefeld interdisciplinary project. Berlin, New York: De Gruyter 571–591.
Sikor T & Müller D (2009): The Limits of State-Led Land Reform: An Introduction. World Development **37**: 1307–1316.
Simon H (1964): Administrative behavior: A study of decision-making processes in administrative organizations. New York: Macmillan.

Simon H (1972): Theories of bounded rationality. Chapter 8. In: Marschak J, McGuire CB, Radner R & Arrow KJ (eds.) Decision and organization: A volume in honor of Jacob Marschak. Amsterdam: North-Holland Pub. Co 161–176.
Simon H (1983): Reason in Human Affairs. Stanford (CA): Stanford University Press.
Simpson SR (1976): Land Law and Registration. Cambridge, New York: Cambridge University Press.
Sjaastad E & Bromley DW (1997): Indigenous Land Rights in Sub-Saharan Africa: Appropriation, Security and Investment Demand. World Development **25**: 549–562.
Sjaastad E & Bromley DW (2000): The Prejudices of Property Rights: On Individualism, Specificity, and Security in Property Regimes. Development Policy Review **18**: 365–389.
Sjaastad E & Cousins B (2009): Formalisation of Land Rights in the South: An Overview. Land Use Policy **26**: 1–9.
Sklenicka P & Hladík JSFKBLJČLŠM (2009): Historical, environmental and socio-economic driving forces on land ownership fragmentation, the land consolidation effect and the project costs. Agric. Econ. – Czech **55**: 571–582.
Skocpol T (1994): Social revolutions in the modern world. Cambridge: Cambridge University Press.
Slider D (1995): Georgia. In: Curtis GE (ed.) Armenia, Azerbaijan, and Georgia. Washington, D.C.: Federal Research Division, Library of Congress 149–230.
Smit B (2002): Atlas.ti for qualitative data analysis. Perspectives in Education **20**: 65–76.
Spivak GC (1988): Can the Subaltern Speak? Basingstoke: Macmillan.
Spoor M (ed.) (2009): The political economy of rural livelihoods in transition economies: Land, peasants and rural poverty in transition. London, New York: Routledge.
Stark D: Not by Design: The Myth of Deisgner Capitalism in Eastern Europe.
Starr P (1988): The meaning of privatization. Yale Law and Policy Review **6**: 6–41.
Stigler GJ (1961): The Economics of Information. Journal of Political Economy **69**: 213–225.
Stiglitz JE (1999): Whither Reform? Ten Years of the Transition, Paper prepared for the Annual Bank Conference on Development Economics, Washington, D.C., April 28-30, 1999.
Stiglitz JE (2002): Whither reform? Towards a new agenda for Latin America: Prebisch Lecture, delivered at the Economic Commission for Latin America and the Caribbean, in Santiago, Chile, on 26 August 2002.
Stiglitz JE (2008): Is there a Post-Washington Consensus Consensus. Chapter 4. In: Serra N & Stiglitz JE (eds.) The Washington Consensus reconsidered: Towards a new global governance. The initiative for policy dialogue series. Oxford, New York: Oxford University Press 41–56.
Strauss AL (1994): Grundlagen qualitativer Sozialforschung: Datenanalyse und Theoriebildung in der empirischen soziologischen Forschung. UTB für Wissenschaft, 1776. München: Fink.
Strauss AL & Corbin JM (1990): Basics of qualitative research: Techniques and procedures for developing grounded theory. London: Sage.
Streeck W & Thelen KA: Introduction: Institutional Change in Advanced Political Economies 3–39.
Streeck W & Thelen KA (eds.) (2005): Beyond continuity: Institutional change in advanced political economies. Oxford, New York: Oxford University Press.
Suny RG (1994): The making of the Georgian nation. 2nd. Bloomington, Indianapolis: Indiana University Press.
Svejnar J (2002): Transition Economies: Performance and Challenges. Journal of Economic Perspectives **16**: 3–28.
Swinnen JFM (1997): Political economy of agrarian reform in Central and Eastern Europe. Aldershot [u.a.]: Ashgate.

Swinnen JFM & Heinegg A (2002): On the political economy of land reforms in the former Soviet Union. J. Int. Dev. **14**: 1019–1031.

Swinnen JFM & Rozelle S (2006): From Marx and Mao to the Market: The Economics and Politics of Agricultural Transition. New York: Oxford University Press.

Szreter S (2002): The state of social capital: Bringing back in power, politics, and history. Theory and Society **31**: 573–621.

Teddlie C & Yu F (2007): Mixed Method Sampling: A typology with examples. Journal of Mixed Method Research **1**: 77–100.

Tesch R (1990): Qualitative research: Analysis types and software tools. London: Routledge-Falmer.

Theesfeld I (2004): Constraints on Collective Action in a Transitional Economy: The Case of Bulgaria's Irrigation Sector. World Development **32**: 251–271.

Theesfeld I, Dufhues T & Buchenrieder G (2017): The effects of rules on local political decision-making processes: How can rules facilitate participation? Policy Sci **3**: 1.

Thiel A (2006): Institutions of Sustainability and multifunctional landscapes: Lessons from the case of the Algarve. ICAR Discussion Paper 13/2006. In: Beckmann, V & Hagedorn, K (eds) Institutional Change in Agriculture and Natural Resources (ICAR). Berlin.

TI Georgia (2010): Property rights in post-revolutionary Georgia. Available from: https://www.transparency.ge/en/content/stub-58.

TI Georgia (2012a): Property in Mestia, Acknowledged Chosen Traditional Ownership. Available from: http://www.transparency.ge/en/blog/property-mestia-acknowledged-chosen-traditional-ownership.

TI Georgia (2012b): Stripped Property Rights in Georgia: Third report. Available from: http://transparency.ge/sites/default/files/post_attachments/StrippedPropertyRights_April2012_Eng_0.pdf.

TI Georgia (2008): Results of TI Georgia's Monitoring and Analysis of the Work of Local Government Bodies. February 5, 2008. Available from: http://www.transparency.ge/files/215_401_684980_TI%20Georgia%20-%20Results%20of%20Monitoring%20and%20Analysis%20of%20the%20Work%20of%20Local%20Governments%20-%20E.pdf.

TI Georgia (2013): Judiciary after parliamentary elections 2012. July 25, 2013. Available from: http://www.transparency.ge/node/3280.

TI Georgia (2014a): Ban on land sales – stories from large foreign farmers. Available from: http://www.transparency.ge/en/node/3968.

TI Georgia (2014b): Government spending priorities in 2012-2014: Analysis and recommendations. February 20, 2014. Available from: http://www.transparency.ge/en/node/3954.

Trebilcock M & Veel P-E (2008): Property rights and development: The contingent case for formalization. Journal of International Law **30**: 397–482.

Trier T (2011): Minority Issues Mainstreaming in the South Caucasus. Available from: http://www.ecmi.de/publications/detail/minority-issues-mainstreaming-in-the-south-caucasus-a-practical-guide-216/.

Tripp AM (2004): Women's Movements, Customary Law, and Land Rights in Africa: The Case of Uganda. African Studies Quarterly **7**: 1–19.

Tsereteli M (2006): Banned In Russia: The Politics Of Georgian Wine. Available from: http://www.cacianalyst.org/newsite/newsite/?q=node/3904.

Tsomaia E/EJ/SD (2003a): The Other Agricultural Land Reform in Georgia: State Leasing of Land to Private Farmers. Available from: http://www.terrainstitute.org/georgia_report/AgLandReform_Georgia.pdf.

Tsomaia E/EJ/SD (2003b): The Other Agricultural Land Reform in Georgia: State Leasing of Land to Private Farmers. Available from: http://www.terrainstitute.org/georgia_report/AgLandReform_Georgia.pdf.

Tullock G (1977): The Paradox of Revolution. Public Choice **Vol. 11**: 89–99.

Tullock GC (1965): The politics of bureaucracy. Washington, D.C.: The Public Affairs Press.

Tullock GC (1967): The Welfare Costs of Tariffs, Monopolies, and Theft. Western Economic Journal **5**: 224–232.

Turmanidze K (2001): State against the invisible: The case of georgian informal economy.

Turmanidze K (2002): Structures of Power in Georgian Municipalities. Available from: www.policy.hu/turmanidze/papers.html).

Tversky A & Kahneman D (1974): Judgment under Uncertainty: Heuristics and Biases. Science **185**: 1124–1131.

USAID (2005): Final Report, USAID Georgia Land Market Development Activity: June 2001 - July 2005. Available from: http://www.aplr.org/files/2/5ew5tjj4ku.pdf.

USAID (2010): Georgia -Property Rights and Resource Governance Profile. Available from: http://usaidlandtenure.net/usaidltprproducts/country-profiles/georgia-1/georgia-country-profile/at_download/file.

USAID (2011a): Land Ownership and Leasing Issue Brief: Unoffical Draft.

USAID (2011b): Evaluation of the Georgia Land Market Development Program.

USAID (2013): Current state of land market in Georgia: Analysis and recommendations. The Economic Policy Research Center. Tbilisi.

Van Huylenbroeck G (2003): Hybrid governance structures to respond to new consumer and citizens' concerns about food. In: Van Huylenbroeck G, Verbeke W, Lauwers L, Vanslembrouck I & D'Haese M (eds.) Importance of policies and institutions for agriculture. Gent: Academica Press 191–206.

Vatn A (2002): Multifunctional agriculture: some consequences for international trade regimes. European Review of Agricultural Economics **29**: 309–327.

Vekua A, Lordkipanidze D, Rightmire GP, Agusti J, Ferring R, Maisuradze G, Mouskhelishvili A, Nioradze M, De Leon, Marcia Ponce, Tappen M, Tvalchrelidze M & Zollikofer C (2002): A new skull of early Homo from Dmanisi, Georgia. Science **297**: 85–89.

Velichkovsky BM (2005): Modularity of cognitive organization: why it is so appealing and why it is wrong. In: Callebaut W & Rasskin-Gutman D (eds.) Modularity: Understanding the development and evolution of natural complex systems. Cambridge, Mass., London: MIT Press 353–383.

Vucinich A (1952): Soviet Economic Institutions: The Social Structure of Production Units. Stanford: Stanford University Press.

Wassmund H (2005): Georgien in der Ära der Sowjetunion - ein Kapital politischer Geschichte. In: Schröder B (ed.) Georgien: Gesellschaft und Religion an der Schwelle Europas. Saarbrücken: Röhrig 9–18.

Waters CPM (2004): Counsel in the Caucasus: Professionalization and law in Georgia. Law in Eastern Europe, no. 54. Leiden, Boston: M. Nijhoff.

Watson RA & Pollack JB (2005): Modular interdependency in complex dynamical systems. Artificial life **11**: 445–457.

Wegren SK (2002): Rural orientations to land privatization in Russia. J. Int. Dev. **14**: 1033–1043.

Weingast BR & Marshall WJ (1988): The Industrial Organization of Congress; or, Why Legislatures, Like Firms, Are Not Organized as Markets. Journal of Political Economy **96**: 132–163.

Werfel F (1978): Die vierzig Tage des Musa Dagh. Berlin/Weimar: Aufbau Verlag.

Werin L (2003): Economic behavior and legal institutions: An introductory survey. River Edge, NJ: World Scientific.

Wheatley J (2003): Group dynamics and institutional change in Georgia: a four-region comparison. Available from: www.oei.fu-berlin.de/en/projekte/cscca/downloads/jw_prop.pdf.

Wieringa S (1994): Women's Interests and Empowerment: Gender Planning Reconsidered. Development & Change **25**: 829–848.

Williams H (1977): Kant's concpet of property. Philosophical Quaterly **27**: 32–40.

Williamson I & Ting L (2001): Land administration and cadastral trends — a framework for re-engineering. Computers, Environment and Urban Systems **25**: 339–366.

Williamson IP (2001): The evolution of modern cadastres: In Proceedings of New Technology for a New Century Conference, FWW2001, Seoul, South Korea, 6-11 May 2001.

Williamson J (1990): What Washington Means by Policy Reform. In: Williamson J (ed.) Latin American Adjustment: How Much Has Happened? Washington: Institute for International Economics 5–20.

Williamson J (2004): The Washington Consensus as Policy Prescription for Development: A lecture in the series "Practitioners of Development" delivered at the World Bank on January 13, 2004. Available from: http://www.iie.com/publications/papers/williamson0204.pdf.

Williamson J (2005): The strange history of the Washington consensus. Journal of Post-Keynesian Economics **Vol. 27**: 195–206.

Williamson J (2008): A Short History of the Washington Consensus. In: Serra N & Stiglitz JE (eds.) The Washington Consensus reconsidered: Towards a new global governance. The initiative for policy dialogue series. Oxford, New York: Oxford University Press 14–30.

Williamson OE (2000a): The New Institutional Economics: Taking Stock, Looking Ahead. Journal of Economic Literature **38**: 595–613.

Williamson OE (1971): The Vertical Integration of Production: Market Failure Considerations. American Economic Review **61**: 112–123.

Williamson OE (1975): Markets and hierarchies, analysis and antitrust implications: A study in the economics of internal organization. New York: Free Press.

Williamson OE (1979): Transaction-Cost Economics: The Governance of Contractual Relations. Journal of Law and Economics **22**: 233–261.

Williamson OE (1981): The Economics of Organization: The Transaction Cost Approach. American Journal of Sociology **87**: 548–577.

Williamson OE (1985): The economic institutions of capitalism. New York, London: The Free Press.

Williamson OE (1991): Comparative Economic Organization: The Analysis of Discrete Structural Alternatives. Administrative Science Quarterly **36**.

Williamson OE (1996): Transaction cost economics and the Carnegie connection. Journal of Economic Behavior & Organization **31**: 149–155.

Williamson OE (2000): The New Institutional Economics: Taking Stock, Looking Ahead. Journal of Economic Literature **38**: 595–613.

Williamson OE (2005): The Economics of Governance. American Economic Review. **95**: 1–18.

Winkler E (1965): Landschaftswandel der Kolchis: Session de la Fédération des Sociétés suisses de géographie, Samedi 25 et dimanche 26 septembre 1965. Verhandlungen der Schweizerischen Naturforschenden **145**: 152–153.

Wood G & Gough I (2006): A Comparative Welfare Regime Approach to Global Social Policy. World Development **34**: 1696–1712.

Wood RC (1958): The New Metropolis: Green Belts, Grass Roots or Gargantua? The American Political Science Review **Vol. 52**: 108–122.

World Bank (2009): Georgia Poverty Assessment: Report No. 44400-GE. Available from: http://www-wds.worldbank.org/external/default/WDSContent-Server/WDSP/IB/2009/04/29/000350881_20090429111740/Rendered/PDF/444000ESW0P1071C0Disclosed041281091.pdf.
World Bank (2012): Georgia: Climate change and agriculture. Country note. Available from: http://siteresources.worldbank.org/GEORGIAEXTN/Resources/CN_Georgia_FINAL.pdf.
World Bank (2014): Registering property in Georgia. Available from: http://www.doingbusiness.org/data/exploreeconomies/georgia/registering-property.
World Bank (2015a): Doing Business 2015: Going Beyond Efficiency: Economy Profile 2015. Georgia. Available from: http://www.doingbusiness.org/~/media/giawb/doing%20business/documents/profiles/country/GEO.pdf.
World Bank (2015b): The Jobs Challenge in the South Caucasus – Georgia. The World Bank, January 12, 2015. Available from: http://www.worldbank.org/en/news/feature/2015/01/12/the-jobs-challenge-in-the-south-caucasus---georgia.
Wren MC & Stults T ([1958] 2009): The Course of Russian History. 5th Edition. Oregon (United States): Wipf & Stock Publishers.
WTO (2009a): I. Economic Environment. WT/TPR/S/224. Available from: http://www.wto.org/english/tratop_e/tpr_e/s224-01_e.doc.
WTO (2009b): Trade Policy Review: Georgia: IV. Trade Policies by Sectors. WT/TPR/S/224. Available from: http://www.wto.org/english/tratop_e/tpr_e/s224-04_e.doc.
Ziegler K (2005): Städtebau in Georgien - vom Sozialismus zur Marktwirtschaft. Dissertation an der Technischen Universität Kaiserslautern. Kaiserslautern.
Zimina RP (1978): The Main Features of the Caucasian Natural Landscapes and Their Conservation, USSR. Arctic and Alpine Research **10**: 479–488.
Znovs A (2011): Actual map of Georgia. In: Discover Georgia, 10/06/2012. Available from: https://discovergeorgiablog.wordpress.com/2012/06/10/actual-map-of-georgia/.

www.ingramcontent.com/pod-product-compliance
Lightning Source LLC
LaVergne TN
LVHW041503170726
843492LV00005B/1355